Estimating Home Building Costs

By W. P. Jackson

Craftsman Book Company
6058 Corte del Cedro, P.O. Box 6500, Carlsbad, CA 92008-0992

Library of Congress Cataloging in Publication Data

Jackson, W.P.
Estimating home building costs

Includes index.
1. Building--Estimates. 2. Dwellings--Design and construction.
I. Title.
TH435.J19 692'.5 81-15305
ISBN 0-910460-80-9 AACR2

Illustrations by W.P. Jackson and Lynn Harvey

Table of Contents

Chapter 1

The Building Site

The building lot, or building site, as it will be referred to in this chapter, may or may not be a part of the house cost. Considerations of site selection are included in this book for speculative builders who buy land and build houses for sale. The cost of the building site is not a construction cost unless the house is built for sale. This chapter will serve as a guide to evaluating site costs, which include purchase price of the site, recording and legal fees, engineering fees for the survey, interest, taxes, liability insurance and other expenses incurred before the house and lot are sold.

All too often the builder is content to build the house without considering serious disadvantages of the site. A wise builder considers many conditions that add to the value of the finished home. When the house has been built and the property is appraised, the appraiser will look for the following:

1. The direction of growth

2. How the property fits into the growth pattern

3. Streets' condition

4. The demographic and economic indicator for the area, such as population, employment, vacancies, rate of growth or decline of area and the reason why

5. Accessibility to schools, churches, shopping centers, recreation areas and public transportation

6. Preservation of trees on the site (trees can add as much as 25% to the appraised land value)

7. Terrain

8. Adequacy of water supply

9. Adequacy of sewage disposal

The Move Back To The City
The move to distant suburbs slowed in the mid - 1970's because of higher transportation costs. Urban areas provide established neighborhoods, shopping variety, nearby medical facilities, and a full range of cultural opportunities lacking in distant suburban tracts. Urban construction means that sewer and water connections are usually cheaper, and streets, walks and utilities are already in.

Building in urban areas means the builder has fewer building sites and may have to build on scattered lots. He will have to look harder to find these sites and will probably have to reconsider land he once passed up: lots in older sections of town, and the neighbor's lot that has been used as a garden for many years. He will probably want to take another look at hillside sites that were not considered desirable. These lots can become valuable building sites, with proper planning.

When shopping for building sites in the city or suburbs, beware of overinflated prices such as those caused by rumors of new industry. Get the facts by checking the record of sale at the county recorders office. Selling prices are usually indicated by the taxes paid on the transfer. Take the legal description of the land to the agency where deeds are recorded. If the legal description and the name of the owner of the land are not known, inquire at the local taxation department. You'll need to know the names and addresses of the surrounding property owners. The agency responsible for property taxes will give you the name and address of the owner and the legal description of the property. For a nominal fee any citizen can check these records. The amount of tax paid indicates the value of the property transferred. The value of the stamps is related to the selling price of the property, and varies with each locality. The personnel in the office where these deeds are recorded will tell you how to compute the selling price of the land.

The selling price of adjacent building sites should not necessarily determine how much the builder pays for his building lots; it is only a guide. But note carefully that a building site with the lowest selling price may be the most expensive after it has been developed. This does not always hold true; sometimes reclaimed lots can be bought and developed into ideal building sites for less than the surrounding lots. Estimate your development costs and to see whether other more expensive sites may actually be more economical.

When looking for buildable lots in or near the city, check for the following:

1. One-family dwellings should not take up more than 40% of the lot. Be

Residential Limited, District R-1

Statement of Intent

This district is composed of low-density residential areas plus certain open areas where similar development appears likely. The regulations for this district are designed to stabilize and protect the essential characteristics of the district, to promote and encourage a suitable environment for family life, and to prohibit all activities of a commercial nature. To these ends, development is limited to relatively low concentration, and permitted uses are limited for the most part to single units. Some additional uses may be allowed, such as schools, parks, churches and specific public facilities, but no home occupations (including room renting) are permitted.

Permitted Uses The R-1 Density Residential District permits:

1. Single-family dwellings
2. Customary general farming, but not the raising of farm animals or poultry
3. Schools
4. Churches
5. Parks and Playgrounds, provided that they are unlighted and a planted buffer strip of trees and hedges is present
6. Accessory buildings such as garages, carports, porches and stoops attached to the main building are considered part of the main building. No accessory building may be closer than three feet to any property line.
7. Public utility stations, provided all lot area requirements of the district are met
8. Church bulletin boards and identification signs which are unlighted and are designed in such manner as not to detract from adjacent residences
9. Off-street parking as required

Area Regulations All buildings and uses, unless specified in the code, should comply with the minimum lot size requirements. There shall be a minimum lot size of 12,000 square feet for each dwelling unit.

Setback Regulations The depth of the front yard shall be:

Type Street	*Setback*
Arterial	60 feet from center line of any street right-of-way
Collector	50 feet from center line of any street right-of-way
All Other	35 feet from center line of any street right-of-way

Frontage The minimum lot width at the setback lines shall be one hundred (100) feet.

Yard Regulations

1. *Side Yard* The minimum side yard for principal structure shall be fifteen (15) feet and the total width of the two required side yards shall be thirty-five (35) feet or more.
2. *Rear Yard* Each principal structure shall have a rear yard of thirty-five (35) feet.

Height Regulations Principal buildings shall not exceed the height of thirty-five (35) feet and accessory buildings shall not exceed the height of fifteen (15) feet.

Maximum Lot Coverage The principal building and all accessory buildings shall not cover more than 40% of the total lot area.

Special Provisions for Corner Lots

1. The side that shall be considered the front side of a corner lot shall be the shorter of the two sides fronting on streets.
2. The sideyard on the side facing the side street shall be thirty-five (35) feet or more for both main and accessory buildings.

Sample Zoning Ordinance

Residential Limited, District R-2

Statement of Intent

This district is composed of a medium density of residential units. The regulations for this district are designed to stabilize and protect the essential characteristics of the district, to promote and encourage, insofar as compatible with the intensity of land use, a suitable environment for family life. This residential district is not completely residential as it includes public and semi-public, institutional and other related uses.

Permitted Uses Within the R-2 district the following uses are permitted:

1. All uses permitted in R-1 provided all other R-2 requirements are observed
2. Single-family dwellings
3. Schools
4. Churches
5. Hospitals
6. Parks and Playgrounds
7. Home occupations, as defined under the special clause
8. Off-street parking - permission for off-street parking must be obtained from the Planning Commission and the City Council

Special Provisions for Home Businesses

1. Home businesses shall be incidental to the use of the dwelling and shall not be the essential residential character of the dwelling.
2. No accessory buildings or outside storage shall be used in connection with the home occupation.
3. No machinery that causes noise or other interference shall be allowed.
4. No internal or external alterations inconsistent with the residential use of the building shall be permitted.
5. Residents of the dwelling only may be engaged in the home occupations.
6. No display of products shall be visible from the street and selling of merchandise or the manufacture of merchandise for sale cannot be the primary function of the home occupation.

Area Regulations All buildings and uses, unless otherwise specified in this code, shall comply with the minimum lot size. There shall be a minimum lot size of 7,500 square feet.

Setback Regulations The depth of the front yard shall be as follows:

Type Street	*Setback*
Arterial	60 feet from center line of any street right-of-way
Collector	50 feet from center line of any street right-of-way
All Others	35 feet from center line of any street right-of-way

Frontage The minimum lot width at the setback lines shall be sixty (60) feet.

Yard Regulations

1. *Side Yard* - The minimum side yard for each main structure shall be ten (10) feet and the width of the two required side yards shall be twenty-five (25) feet.
2. *Rear Yard* - Each main structure shall have a rear yard of twenty-five (25) feet or more.

Height Regulations

1. All dwelling units shall not exceed the height of thirty-five (35) feet and all accessory buildings shall not exceed the height of fifteen (15) feet.
2. All public or semi-public buildings shall not exceed the height of sixty (60) feet and rear yard shall be increased one (1) foot for each foot in height over thirty-five feet.

Lot Coverage The principal building and all accessory buildings shall not cover more than forty (40) percent of the total area.

Special Provisions for Corner Lots The same provisions which cover R-1 district for corner lots are observed in R-2 districts.

Sample Zoning Ordinance

sure the lot is of a size and shape so the house planned will fit without violating any local building regulations.

2. The building site should be large enough so that at least one of the front, rear or side areas can be used for drying clothes, landscaping, a driveway and outdoor living.

3. There should be easy access to and circulation around the building. Can the house be maintained without trespassing on adjoining property?

4. There should be sufficient room on the property to assure a safe and sanitary installation for individual water supply and sewage disposal system if needed.

Zoning ordinances vary from community to community, but they are all written to protect the health and safety of the occupants and to ensure the structural soundness of the buildings. Listed on the two prior pages is the building code for one city showing the restrictions for two different zoned residential districts, the Residential Limited, District R-1 and the Residential Limited, District R-2. The zoning requirements shown here are probably similar to the code where you build.

Check Before Buying The Building Site

When a builder buys a site to build on, he hopes to sell the house and lot with a minimum of delay. It is crucial that you consider carefully all the elements of the site before you buy.

When you look for building sites, have a general idea of the type and price range of the houses you want to build. If your houses are cheaper or more expensive than the average house in the neighborhood, you could have a sales problem. Few people want to live in the cheapest or the most expensive house on the block. A safe rule is to keep your house within 15 to 25 percent of the average price in the neighborhood.

Location is an important factor in the value of residential houses. The neighborhood affects the cost of the land which ultimately affects the cost of the house that is built on it. Buy in growing areas. Proximity to schools, shopping centers, public transportation, hospitals and recreation areas is very desirable. Check vacant houses in the neighborhood. If there are several vacant houses in the area, it may be caused by a breakdown in zoning regulations and deed restrictions.

Before purchasing the building site, it is always wise to check zoning regulations with the local planning commission. The regulations may not be enforced. Also check with them on any future plans they may know about that affect your decision to buy the land. A building moratorium may be in effect or may be planned in the near future. A no-growth area can be disastrous financially. Idle land is a 100% liability. Taxes, interest, liability insurance and maintenance are constant expenses that will have to be added to the cost of the building site. If there is a moratorium in the area, delay buying the land until you have assurance from the proper authority that the moratorium will be lifted in the near future. Take options on the land if necessary.

If you are buying for the future and not the present, you should have the answers to the following questions before buying:

1. Can you afford to pay the taxes, interest and other assessments on the land until a house is built and the property is sold?

2. Is the area growing in the direction where you want to buy? If so, is the growth residential, industrial or commercial?

3. Are any highways planned through this area? Check with the highway department for this information.

4. What are the present zoning regulations? What future plans does the local governing body have for the area? The local planning commission will have this information.

There are few areas where there are enough building sites available to select the lot of your choice and orient the house on the land to obtain maximum energy efficiency and maximum livability. Some of the following types of building lots may be worth taking a second look at:

Wooded areas are expensive to build on, but with the proper planning they can become beautiful building sites. They will have a much higher resale value when developed.

Bare land, or land with few trees is cheaper to develop and build on, but landscaping is much more expensive.

Hillside lots usually have a view that can never be taken away. It can be developed into beautiful sites. Some of the most expensive houses built today are built on these lots.

When you have found the land and decided to buy it:

1. Obtain a reputable attorney.

2. Check the courthouse records to verify if a clear deed can be obtained.

3. Check the present and future zoning regulations for the area.

4. Check with the highway department to see if there is a possibility of a highway being built in the foreseeable future on or near the property.

5. Check the present and projected tax rate.

6. Check with the utility companies for getting service lines to the property if they are not there.

7. Have a thorough knowledge of the building codes, deed restrictions, easement rights and any other building requirements for the land.

The cost estimate sheet for the building site is designed to tell the builder what the total cost of the land will be by the time he sells it. Costs other than the site are included in your land cost. Some of these additional costs are explained in the following paragraphs.

Recording and Legal Fees

In many states a lawyer is required in all real estate transactions. A deed

SWIMMING

MACHINE

2
6

426 6644
CREDIT

- GUITAR STAND
HARDWOOD W FELT
- ROLL CAGE CONSTRUCTION
- TRIANGLE STRENGTH
MODERN G FATHER CLOCK
TIGHTWAD GAZETTE
LEEDS, 04267
WELFARE TESTS ME

WT
TEMP
TIME

MCLAUGLIN

MIRROR / 729 5113

must be prepared and recorded. The seller (grantor) normally pays for the cost of preparing the deed, but the buyer (grantee) bears the expense of transferring the property. The grantee will have to pay for the deed of trust (if there is one), the recording fee, his share of the transfer tax and the title check.

The title should be checked for any mortgages, mechanics liens, easements, or other encumbrances that may influence your decision to buy the land.

If the building site is on a private road, find out what assessments will be made for maintenance, snow removal, and utility lines. These expenses will have to be paid and they will add to the cost of the land. If the building site is on a private street, it should be protected by a permanent easement.

The attorney should check the present and projected tax rates, and can help you obtain a thorough knowledge of the zoning restrictions and other building requirements.

Engineering Fees

If the lot corners are missing, engage a registered land surveyor to survey the lot. Make sure the lot described in the deed is the lot you think you bought. There are many instances of people building on the incorrect lot. If this happens the building will have to be moved or the building site purchased at whatever price the owner demands.

The engineer can advise you on conditions on the lot that will have a direct bearing on the construction costs. An example of such a condition would be topography of the land as it affects the grade of the driveway, walks and drainage. Plans for the outdoor living area may have to be revised because of the topography of the land. The building site must be free of hazards that may affect the health and safety of the occupants, or the structural soundness of the building. Such hazards include subsidence (excessive settlement from unstable soil, high ground water and springs), flood, and erosion. Underground springs cause hydrostatic pressure which results in leaks in basement floors. High ground water will cause the additional expense of waterproofing the basement, involving weeping tile, sump pumps, and waterproofing foundation walls. Springs and high ground water may require raising the elevation of the finished floor.

Run-off water may cause damage to the surrounding property. Many zoning regulations prohibit the construction or erection of any building or structure within the city unless there is in force an approved erosion and sedimentation plan. It may also be unlawful for any person to clear, grade, excavate, fill, remove topsoil from or change the contour of any land in the city unless there is an approved erosion and sedimentation plan. It may also make it unlawful to remove or destroy trees, shrubs, or other plant life without prior approval. Here again the engineer can advise you. One helpful note can save the builder some money: If storm sewers are required use the smallest size permissible. Any decrease in the size of the storm pipe is an increase in savings in development costs.

When an on-site sewage disposal system is required the lot must be large enough to meet the local sanitation requirements. The soil must be tested for water absorption (percolation tests), a job for the engineer.

When the engineer locates the corners of the lot he should use permanent markers rather than wooden stakes. This may cost a little extra money now but it may save the expense of relocating the corners later because the wooden stakes were knocked down and lost.

Cost Estimate Work Sheet For Building Sites

1. Purchase price of site $________
2. Recording and legal fees ________
3. Engineering fees ________
4. Interest ________
5. Taxes ________
6. Maintenance ________
   (Clearing underbrush, etc. while lot is idle)
7. Other costs ________
   (Liability insurance, assessments, etc.)

(*) Total cost of site $________
(Enter on Line 1, Form 100)

(*) Form 100 is the Cost Estimate Form used to compute the total cost of the house and building site - it is shown in detail in the last chapter of this book. If the cost of the building site is not to be included in the construction costs then enter zero on Line 1, Form 100.

Chapter 2

Preliminary Costs

The architect's fee (if one is engaged to draw the blueprints), or the cost of the stock plans may or may not be part of the construction costs for the builder. For the merchant or speculative builder who builds houses for sale, they will be part of his construction costs. However, if he is bidding on a job on a contract or cost-plus basis, these blueprints will be paid for by the owner and this fee will not be included in the cost estimatc.

Written instructions for the construction of even the smallest building would, if not accompanied by drawings, be impossible for anyone to understand. These drawings are called blueprints, or plans because they used to consist of white lines on paper with a blue background. This paper was sensitized (coated) and when it was run through a blueprint machine with the tracing paper over strong lights it was sprayed with a solution and then passed over heaters to dry. The sensitized paper would turn a deep blue color except where the lines on the drawing paper protected it from the light. Those lines would remain white. The reverse method of using sensitive paper that reproduces dark lines on a white background is used today. This method is preferred by builders because it is easier to make notes on the white background. Either type of paper is still referred to as a blueprint.

Except in the case of large tract or very expensive homes, an architect is not normally engaged to draw the blueprints for residential construction. Stock plans can be bought from the many business firms that specialize in home planning services for a nominal fee. Plan book companies offer plans for houses that vary widely in size, design and livability. Most people who are planning to build a house will have little trouble finding a stock plan that (perhaps with a few minor changes) is compatible with their needs and taste. If the house is to be a factory-built house, the manufacturer will furnish all necessary blueprints as a part of the package price.

There are few plans that are so complete that the builder does not have to rely on his technical training and expertise to recognize any necessary changes. In the interest of good workmanship and economy a thorough set of plans should be required. Unless a person is proficient in drawing house plans, hire a professional to do it. There are many cases of amateurs drawing their own plans to save the cost of the blueprints. It is unnecessary to say what an error in the blueprint does to the cost of constructing a house.

Blueprints are necessary not only for the builder and the subcontractor, but also in order to obtain building permits and financing. Architectural and engineering exhibits must be submitted with the application for mortgage loans. These exhibits must describe precisely the size and location of the house and the quality of materials and equipment to be used. Before the loan is finalized an appraisal will be made and here faulty workmanship (which may be due to incomplete plans) can have an adverse effect on the appraised value of the property.

Plot Plans

Plot plans are a vital part of the working drawings used in building a house. They are essential for the builder or estimator in estimating the construction costs. Plot plans may be prepared by the builder or an engineer. If they are prepared by an engineer this fee will probably be included in the engineering costs of the building site as explained in Chapter 1. Normally, the cost of the plot plans is paid by the owner of the property. If the job is let out on bids this fee would not be part of the construction costs for the builder.

The plot plan must show the following minimum information:

1. Compass direction showing north
2. Metes and bounds of the property lines and their distances
3. Lot corners
4. Description of the lot and section number
5. Elevations of each different floor level
6. Location and dimensions of easements
7. Location of the house on the lot
8. Location of existing trees
9. Grade elevations

10. Location and dimensions of water and sewer lines, and the location of electric, gas, telephone and T.V. cable service

11. Finished floor elevations of the dwelling, garage or carport, street where the driveway connects, and the grade elevations at each corner of the dwelling

12. Location and dimensions of the house, garage, carport and other buildings

13. Location and dimensions of walks and the driveways

14. Location of steps, terraces and porches

15. Scale of the drawing

The elevations can be given in feet above sea level (if they are known) or above or below an arbitrary elevation based on a fixed point called a bench mark or B.M. If an arbitrary elevation is used, the bench mark is normally assigned an elevation of 100.0'. All other elevations on the plot plan are in reference to this point. The bench mark may or may not be shown on the plot plan.

The accompanying plot plan (Figure 2-1) is for a two-story house with a basement and a two-car garage. From the information given on it the builder or estimator can make the following calculations for his construction costs:

1. From the metes and bounds of the lot the builder can calculate the interior angles of the lot corners. A book titled *Building Layout,* available from Craftsman Book Company (6058 Corte del Cedro, P.O. Box 6500, Carlsbad, CA. 92008) shows the reader how to convert bearings to interior angles. It is important for the builder to know this because frequently one or more of the lot corners are lost, or there may be an obstruction, such as trees, or rocks, that prevents sighting from one lot corner to another with the transit. All lot corners are not 90 degrees and when the building restrictions require a minimum setback from the property line, a violation may be very costly.

2. The finished floor elevations and the grade elevations tell the estimator if extra material will be required and how much excavation or fill dirt will be needed. As an example, in the accompanying plot plan the elevations show that an estimated 1650 cubic yards of compacted fill dirt will have to be imported to the jobsite. The basement floor elevation shown (96.23') is lower than the inverse elevation of the sewer line at the manhole (97.12') where the sewer connection will be made. Although there will be no bathroom or laundry facilities in the basement a sump pump will be required to discharge any water that may get in the basement.

3. The location of the building and the dimensions of the driveway and walks determine their cost.

4. The location of the utility connections will show how far the service lines must be run.

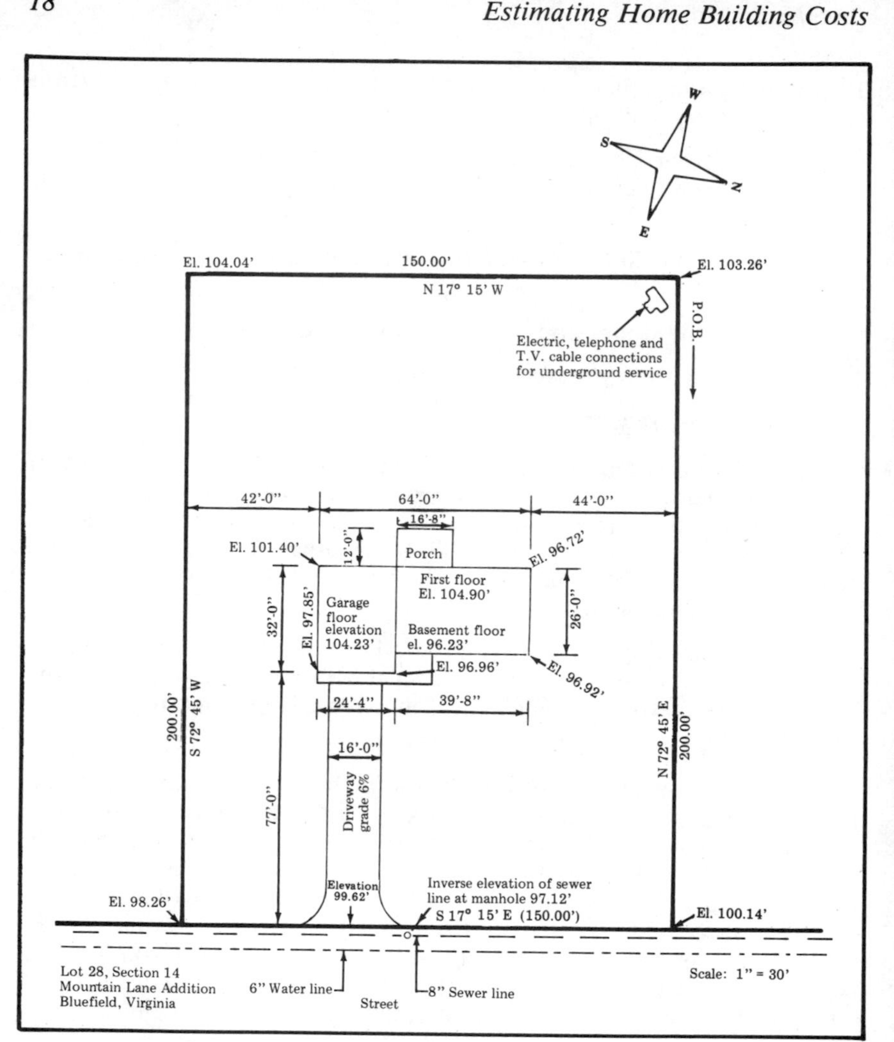

Plot Plan
Figure 2-1

5. Any easement rights shown on the plot plan may change the plans that may be made for that particular area, such as accessory buildings or outdoor living areas.

Building Permit

The building permit is an authorization by the building department of the governing body with jurisdiction over the area to construct a new building, or to alter or enlarge an existing unit. The governing body will be the city or town if the building is to be built within the corporate limits, or the county if the construction is outside of the city limits.

BUILDING PERMIT

ALL SPACES MUST BE COMPLETED

Date April 17, 1980

A BUILDING PERMIT MUST BE ISSUED BEFORE STARTING CONSTRUCTION.

Application for a Building Permit must be made to the Building Officials.
Application is hereby made for a Building Permit in accordance with the description and for the purpose hereinafter set forth. This application is made subject to all local and State laws and ordinances and which are hereby agreed to by the undersigned and which shall be deemed a condition entering into the exercise of this permit.

Name of Owner W.P. Jackson Address 109 Fincastle Lane

Name of Contractor Builder Same Address

Certified State Contractor's No. 15474 Zone Classification R-1

If for Alteration or Repairs, state in detail NONE

*Zone classification must be specified.

Name of Street Mt. View Court Lot No. 28 Section of Mt. View Subdivision.

Size of Lot 150.0' x 200.0' If purchased within the past two years from N/A

Date I hereby certify that on January 1 the land described above is listed in the name of W.P. Jackson

NOTE: Permit for septic tank and approval of location and of well must be obtained from the County Health department after lot has been cleared and building has been staked out, but before construction has been started.

Plot Sketch N/A Construction Plans N/A

Estimated date of completion September 15, 1980

I hereby certify that I have the authority to make the foregoing application, that the information given is correct and that the construction will conform with the regulations in the Building Codes, Zoning Ordinances and private building restrictions, if any, which may be imposed upon the above property by deed. I also agree to be responsible for any and all damage to any and all property, public or private, caused by above construction/repair.

Signature of owner or authorized agent

Address 109 Fincastle Lane Phone Number 326 0539

I, the undersigned, hereby make application for a permit to ~~alter~~/erect a two-story building to be used for residence (Commercial/Residence) on my property located at 214 Street, Mt. View Court Lot No. 28 Section of 14

Sub-division Mt. View

The general shape of my lot and the location of the proposed improvements are accurately set forth in plot plan, on the other page.

Front yard available 77'0" ft. Type of heating Electric

Side yard available 42'0" ft. Number of flues None

Side yard available 44'0" ft. Size of flues N/A

Note: All flues must be lined.

Rear yard available 91'0" ft.

Type of roof Asphalt shingles Ceiling joists and rafters on 16 in. centers.

Expected amount of total electrical load 200 amps

Type of floors Wall to wall carpet over sub-floor and floor underlayment Joists on 16" centers.

No. of rooms Seven Type of construction Drywall and panel

No. of baths Three Type of construction Drywall

Basement size 39'8" x 26'0" Type of construction Concrete block with concrete basement floor

Garage/carport 24'4" x 22'0" plus 7'6" x 10'0" offset Square feet 610

First floor, square feet 1229 Second floor, square feet 1040

Note: Soil bearing test for other than one family dwelling shall be required. Report of soil test N/A

I, the undersigned, do affirm that all the foregoing figures and statements are true, full and correct to the best of my knowledge and belief, and all sanitary, safety and building ordinances of the Town of Bluefield, Virginia will be complied with in said construction. I tender with this application the sum of $77.09 covering permit, $200.00, covering sewer tap, and $200.00 covering water tap.

A TOTAL AMOUNT OF $477.09

Building Permit Approved ....

Disapproved ....

Applicant W.P. Jackson

Code Reference ....

By:

Building Inspector ....

Engineer ....

Manager ....

Building permits regulate the location, size and type of buildings that can be built in a particular area. The building inspector uses the blueprints and plot plan to make sure that there is no violation of the zoning ordinance for the area. If, after checking the blueprints and plot plan, he is satisfied that the proposed building does not violate the zoning restrictions, he may issue a building permit. If a minor violation is present, the building inspector will present the application for the building permit to the Planning Commission to be approved or disapproved. If the violation is not serious, they often will overlook the violation and recommend issuance of the permit.

In addition to the building permit the owner, or builder, must obtain a plumber's permit and in most cities, an electrical permit. Because the plumbing and electrical work is done by subcontractors, the responsibility and expense of obtaining these permits is theirs. This cost will be included in their bid and is included in the construction costs under their respective heading.

The cost of building permits varies among localities. One common cost basis is the value of the structure. The rate may be based on a set fee for the first determined valuation and another fee for each additional thousand dollars. For example, the fee for the building permit may be $45.00 for the first $15,000.00 of valuation and $2.00 for each additional $1,000.00. An application for a permit to build a house with an estimated value of $75,000.00 would cost $165.00. The cost of the permit is computed as follows:

Cost of Permit		Valuation
$ 45.00	for first	$15,000.00
120.00	for next	60,000.00
$165.00	on valuation of	$75,000.00

Obviously, the value of any property is debatable. Many disputes have arisen between the property owner and the governing body issuing the building permit over its cost when the valuation guideline is used. For this reason many cities now base the cost of the permit on the square feet of the house, as shown in the example below:

$.03 per sq. ft.	First floor
$.01½ per sq. ft.	Second floor
$.01½ per sq. ft.	Garage
$.01½ per sq. ft.	Basement

Thus the cost of the building permit would be:

First floor	2100	sq. ft. @ $.03	=	$ 63.00
Second floor	1200	sq. ft. @ $.01½	=	18.00
Garage	680	sq. ft. @ $.01½	=	10.20
Basement	1040	sq. ft. @ $.01½	=	15.60
	Cost of building permit			$106.80

A typical application for a building permit for a two-story house follows. This application includes the plot plan in Figure 2-1. The blueprints submitted with the application may or may not be returned.

Water Connection

When the water supply is provided by a public utility there will be a fee charged for the connection. The service line may be brought to the building site, or the owner or builder may have to run the line from the building site to the main line. Find this out before you make a cost estimate of the house.

If there is no public water supply available, you will have to dig a well, or have water piped from springs that may serve more than one family. If the water supply is privately owned have the water checked for purity by the local health department. Be sure you know who is responsible for maintaining this private water supply.

You must know the total cost of getting an adequate water supply to the building site, before beginning the cost estimate.

Sewer Connection

As with the water supply, the sewage disposal may be a public sewer line where you will be charged for the sewer tap, or it may be a private sewage disposal system. This private system may be a septic tank or a privately owned sewage disposal system that serves more than one family. If a septic tank must be installed, obtain a permit first from the local health department. When the owner or builder submits an application for the permit, an agent from the health department will come to the building site and design the sewage disposal system for the proposed house. Factors that will be used to determine the size of the septic tank include findings from percolation tests, the size and shape of the building site, the size of the house and whether or not it will have a garbage disposal. After the system is designed by the health department they will issue a permit to build it. Before the system is covered over, an inspector will check it for compliance before final approval is given.

These expenses will be part of the construction costs and must be included in the cost estimate.

Temporary Water Service

If the water supply is from a public utility, estimate the cost of the water used during the construction of the house and figure it in the cost estimate. In most cases, there will be a minimum charge per month for the water. Except for brief periods, such as when the brick masons or dry wall subcontractors are working, the minimum monthly charge will cover the cost of the water used. Normally, the minimum monthly charge for the water multiplied by the estimated number of months required to complete the house will give a fairly accurate estimate of the cost of the temporary water service. The cost estimate work sheet at the end of this chapter shows how to compute this cost.

Temporary Electric Service

Because the permanent electric service cannot be installed until the house is built, a temporary meter will be required for service during the construction. The electric utility company will charge a fee for installing this meter as well as a minimum monthly fee for the electricity used. As with the temporary water service, the minimum monthly charge for the electricity, multiplied by the estimated number of months to complete the house, will give a fairly good estimate of the cost of electricity to be used. A note of caution should be made here: in cold weather add an allowance for the electricity to run any needed heating units. The cost estimate worksheet at the end of this chapter shows how to compute the total cost of electric service.

Cost Estimate Work Sheet For Preliminary Costs

2.1 Architects fee or cost of plans $_________

2.2 Plot plans _________

2.3 Building Permit:
First Floor ____sq. ft. @ ____ = $_________
Second Floor____sq. ft. @ ____ = _________
Garage ____sq. ft. @ ____ = _________
Basement ____sq. ft. @ ____ = _________
Total $_________ $_________

If Cost Of Permit Is Based On Value
$_________ @ $_________ = $_________
$_________ @ $_________ = $_________
$_________ @ $_________ = $_________
Total $_________ _________

2.4 Water connection _________

2.5 Sewer connection _________

2.6 Temporary water service:
(a) Minimum cost per month $_________
(b) Estimated months to complete $_________
(a)____ x (b) ____ = $_________ _________

2.7 Temporary electric service:
(a) Charge to install temporary meter $_________
(b) Minimum cost per month _________
(c) Estimated months to complete _________
(a) $_________
(b) $____ x (c)____ _________
Total $_________ _________

Total preliminary costs $ _________

(*) (Enter on Line 2, Form 100)

(*) Form 100 is shown in detail in the last chapter.

Chapter 3

Site Clearing, Excavation and Fill Dirt

Residential construction is a highly competitive business. The success or failure of any builder begins with the accuracy of his cost estimate for each individual job. This is as true for the merchant, or speculative builder as it is with the builder that is bidding on a contract. The merchant builder will arrange for his financing on the basis of the estimated cost of the house. If he sells the house before he knows his total construction costs, and has underestimated by a substantial amount, he may lose money. If he has underestimated the cost of the house he may need additional funds to complete it, and this will harm his banker's confidence in him.

Bad estimates cost money. If the estimate is too low the builder will lose money; if it is too high he probably won't get the job. Do not get careless on a hurry-up estimate and omit something. Once a bid is submitted and a contract is signed it is binding. There are many cases in which the builder or estimator omitted a major item, such as the brick on a brick house and didn't discover the error until it was too late. Such an error can be financially disastrous.

Never make a firm and binding estimate on the cost of a house by the square foot, or cubic foot method. It is not accurate enough and should only be used to get a "ball park" figure. For example, if a house has 2000 square feet of living area and has a rectangular design with four exterior corners, it can be built cheaper per square foot than another house that has the same number of square feet, but has eight exterior corners. The type and quality of the material that is used in the house, and the pitch of the roof can change the cost of the house by several thousand dollars and not add one square foot to its size. Note also that the same house plan and specifications built with the same crew of workmen on adjacent lots can still vary in price because of the development costs. Estimate the cost of each house separately.

Costs can vary for identical items of work. Inflation today often increases costs from week to week. Labor productivity varies with each different journeyman, and with each different day. It may be up one day and down the next depending on his mental attitude and how he feels physically. Here is where experience and accurate cost records from previous jobs are essential. Make allowances in any cost estimate for conditions that vary from normal. In addition to inflation pilferage, cold weather, and lost time have to be reckoned with in any construction estimate.

A day lost during construction adds a day to the time it takes to complete the house. Each extra day in completing the job adds to the interest costs, and this comes when the interest is being paid on the full amount of the construction loan. Delays do happen and if they are kept to a minimum the expense of these delays can be estimated in advance with reasonable accuracy. To help avoid these delays schedule the work on a day-to-day basis. Enlist the cooperation of the supplier and subcontractors. The Critical Path Method, a graphical method used to conrol the planning and scheduling of a construction job to minimize lost time, is in common use among builders today.

On any estimate for residential construction the actual cost should be within 5% of the estimated cost. If you subcontract most of the labor and use a step-by-step cost estimate to prevent errors of omission, you can obtain this accuracy.

Site Clearing

The building site must be cleared and made ready before construction can begin. Always visit the site before making your cost estimate.

Some building sites are wooded or have old buildings and paved areas that must be removed. Old fences and poles, debris of old brick, blocks, brush and tree limbs may be present and should be removed. Wood buried underground may cause termites and should always be burned or removed from the building site.

Small trees and stumps can be removed by a dozer or loader, but the larger trees will have to be cut down by an experienced tree trimmer. The loader can demolish old fences, poles and paved areas, but any buildings on the site that must be demolished should be done by hand. There may be some usable material in the building, and it would be destroyed if a dozer or loader were used.

It is difficult to estimate the site clearing costs when there is a considerable amount. Ask contractors who specialize in this work to bid on it.

The following costs may be used as a guideline:

Removal of trees	
Diameter from 12" to 36"	$40.00 to $60.00 each
Stump removal	
Diameter from 12" to 36"	$100.00 to $200.00 each
Removal of paved areas	
Bituminous paving	$1.00 to $2.00 per square yard
Concrete paving	$2.00 to $3.00 per square yard
Removal of old buildings	
Depending on size of building, allow	$1.00 to $2.00 per square foot
Removal of old fences	
Depending on type of fence, allow	.25¢ to .50¢ per linear foot

Removal of debris from site

Depends on the quantity of debris and the distance it has to be hauled. Check with the owner of the equipment for a firm price.

The above costs are only approximate and should not be used as a firm cost basis. When the site clearing costs will amount to a sizable expenditure of money, get a firm bid on the work before completing the estimate.

Excavation

Estimating excavations is difficult. More errors are made here than in any other phase of residential construction. The construction of any foundation will require the removal of some earth. The quantity of earth to be removed depends on the design of the house, the floor elevations, the location on the building site, and if the house has a basement, crawl space, slab or a combination of these types of foundations.

The builder or estimator should have some knowledge of taking elevations, calculating floor elevations and computing the cuts and fills. If he does not have any experience in this work he should consult an engineer. A book *Building Layout*, published by Craftsman Book Company, 6058 Corte Del Cedro, P.O. Box 6500, Carlsbad, CA. 92008, can be very helpful in this. The builder or estimator must know the answers to the following questions for the excavation estimate:

1. Is the building site large enough to store the soil during the construction period, or will it have to be hauled away? If it has to be hauled away, how far will it have to be hauled, and will it have to be hauled back?

2. Will shoring be required?

3. Will safety precautions be required around the excavated area (in high density areas some cities require it)?

4. Will blasting be necessary?

5. Has there been a land fill in recent years that may require additional excavation, or piling to reach solid earth?

There is a difference between a general contractor and a residential builder. The general contractor contracts for earth moving jobs, water and sewer lines, commercial and industrial buildings. He must have the earth moving equipment necessary for this work. The small volume residential builder normally subcontracts the excavation for his jobs. When estimating the excavation costs, make an allowance for the soil conditions, hauling the equipment to the building site, and if there is rock present. Here again, accurate cost records from previous jobs are very helpful. Machine excavation where there is no rock should cost between $1.00 to $2.00 per cubic yard depending on the soil conditions and the location. If blasting will be required it can increase the excavation costs by 200% or more, depending on the severity of the blasting and proximity of the surrounding buildings and utilities.

Working in the winter is another factor the estimator must reckon with. Cold weather can increase excavation costs by 15% or more. Frozen earth is much harder to excavate and haul and damages the earth moving equipment. A wet weather or freeze-thaw cycle is also troublesome as the top soil is soft and muddy and the remaining earth is still frozen. Builders in some areas try to get all rough grading done before winter but this adds to the interest paid on borrowed funds until work can be started up in the spring. The access roads can be graded, compaction completed and graveled so trucks can deliver without getting stuck in the mud. These costs must not be overlooked in the cost estimate. It is always a good idea to rough grade the site so surface water will flow away from the foundation. This may save many days of lost time. The cost of preparing the area for storing and protecting the material should not be overlooked.

Figure 3-1 is the foundation plan of a house that has been staked out on the building site from the plot plan. The existing grade elevations at each corner of the house have been taken and the floor and grade elevations have been computed and recorded. From this information the calculations can be made for the amount of earth that will have to be excavated. Allowing for the excavation to extend two feet outside of the building line around the perimeter of the foundation, the number of cubic yards of earth that will have to be excavated is computed as follows:

1. The dimensions of the basement foundation wall from the blueprints and shown on the plot plan are 39'8'' x 26'0''. Allowing for the two additional feet for working around the foundation, the basement excavation dimensions will be 43'8'' x 30'0''.

2. The basement grade elevation has been computed and is 95.57'. The grade elevation at the right back corner of the foundation is 96.72' and a cut of 1.15' (1'2'') will be required here [95.57' less 96.72' = - 1.15'].

3. The grade elevation at the right front corner of the foundation is 96.92'. A cut of 1.35' (1'4'') will be required here [95.57' less 96.92' = -1.35'].

4. The left front corner of the foundation adjacent to the garage area has a grade elevation of 96.68'. Here a cut of 1.11' (1'1'') will be required [95.57'] less 96.68' = - 1.11'].

5. The left back corner of the foundation adjacent to the garage area has

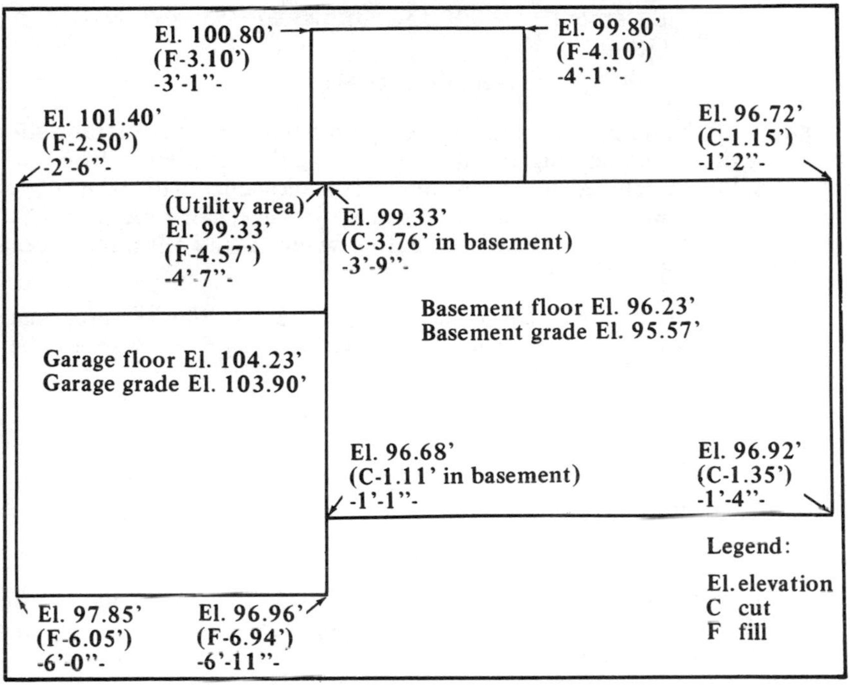

El. 100.80'
(F-3.10')
-3'-1"-

El. 99.80'
(F-4.10')
-4'-1"-

El. 101.40'
(F-2.50')
-2'-6"-

El. 96.72'
(C-1.15')
-1'-2"-

(Utility area)
El. 99.33'
(F-4.57')
-4'-7"-

El. 99.33'
(C-3.76' in basement)
-3'-9"-

Basement floor El. 96.23'
Basement grade El. 95.57'

Garage floor El. 104.23'
Garage grade El. 103.90'

El. 96.68'
(C-1.11' in basement)
-1'-1"-

El. 96.92'
(C-1.35')
-1'-4"-

Legend:
El. elevation
C cut
F fill

El. 97.85'
(F-6.05')
-6'-0"-

El. 96.96'
(F-6.94')
-6'-11"-

Foundation Plan
Figure 3-1

a grade elevation of 99.33'. This will require a cut of 3.76' (3'9") [95.57' less 99.33' = - 3.76'].

The excavation for the basement area is the only excavation that will be required for this house foundation. The other areas of the foundation will require fill dirt, which willl be explained later in this chapter. To compute the cubic yards of earth that will have to be excavated for the foundation shown in Figure 3-1 do the following:

1. Average the depth of the cuts for the basement area. These cuts average 1.84' (1'10").

2. Change 43'8" (the length of the basement foundation) to feet and its decimal equivalent of a foot. (43'8" is 43.67').

3. The number of cubic yards to be excavated is:

$$\frac{43.67' \times 30.0' \times 1.84'}{27} = 89.28 \text{ cubic yards}$$

To provide an access road and an area for storing the material, an additional 65 cubic yards of earth will have to be excavated. The total cubic yards of the excavation will be:

Basement	89.28 rounded to	90
Other		65
	Total cubic yards	155

There will be no problem storing the earth from the excavation on the building site, so no hauling expense for moving the earth away from the site should be included in the estimate. No allowance will have to be made for blasting as there is no rock present in the excavation area. The house is scheduled to be completed before winter so no allowance need be made for cold weather.

Allowing for the cost of moving the equipment to the job site, past cost records, and allowing for inflation, $1.45 per cubic yard should be the approximate cost of the excavation.

$1.45 x 155 cubic yards = $224.75

The cost of the excavation will be estimated at $225.00 (rounded off from $224.75).

Fill Dirt

As in estimating excavations, many errors are made in estimating fill dirt. The quantity of fill dirt that will be required may be more hazardous to an estimator than estimating the quantity of earth that has to be excavated. The compaction will add approximately 25% to the total cubic yards that will be needed. The present grade elevations, the finished grade elevations and the location of the dwelling on the building site are factors in estimating the quantity of fill dirt. The elevations shown on the plot plan are essential in making these calculations.

Volume of loose dirt can be as much as 25% greater than the original "in place" volume. Uneven settling can take place for many months, or years, depending on the depth of the fill, the compaction of the soil and the type of soil used. Cinders, sand, gravel and shale are suitable in some of the deeper fill areas, but it should never be used where it will affect the landscaping. Beautifully landscaped yards can be ruined by deep and uneven settling.

Never use organic material in land fills. The settling will take place for years. It can also be a very costly mistake to use organic material as fills under a slab.

Compaction of the fill dirt should be done in layers. On deep fills the compaction should be continuous as the fill dirt is imported to the area. Fills that support slabs should be compacted in 4" to 6" layers.

From the grade and floor elevations, and dimensions as shown in Figure 3-2, and the enlargement of the floor plan in Figure 3-3 showing the cuts and fills, the following computations can be made:

1. The dimensions of the garage and utility area are 24'4" x 32'0". A fill will be required in the entire area and it must be compacted in 4" to 6" layers, as it will support the concrete floor.

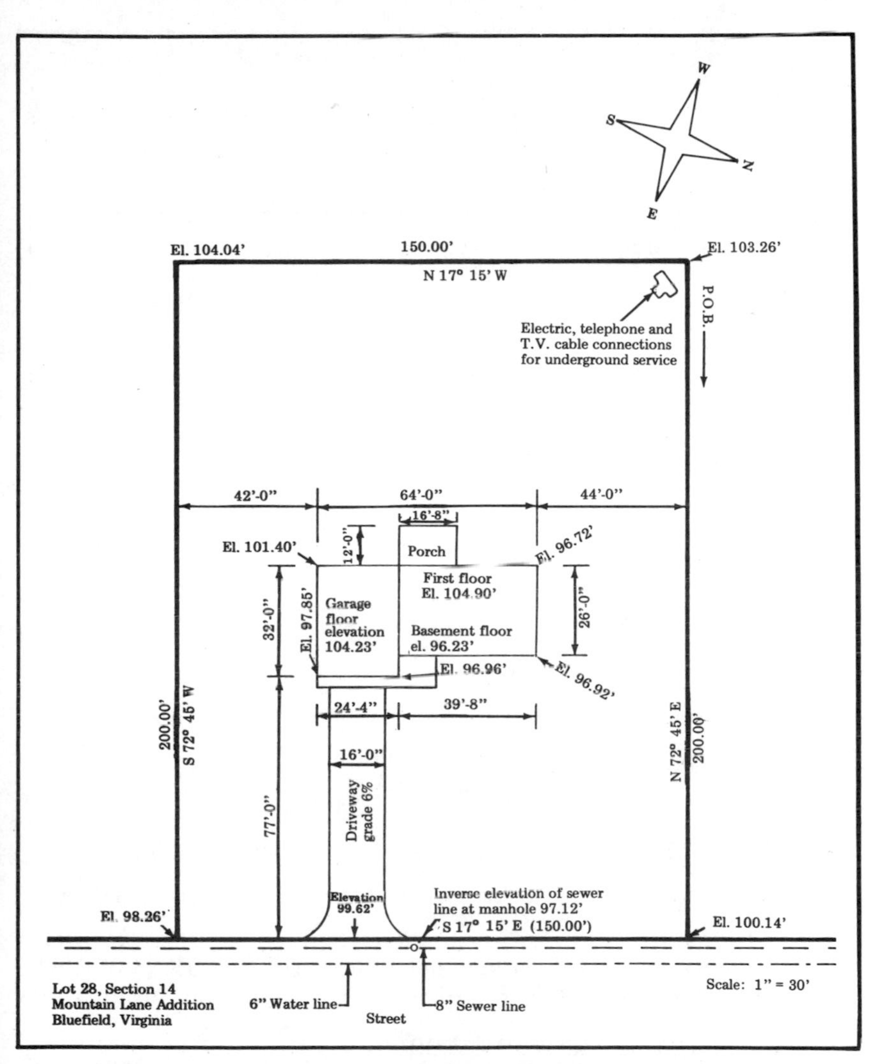

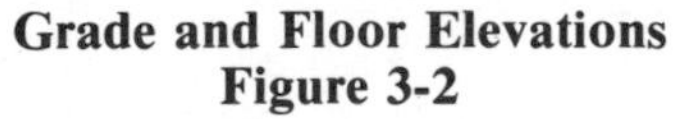
Grade and Floor Elevations
Figure 3-2

2. The garage and utility floor elevation has been calculated to be 104.23'. The specifications call for a 4" concrete slab over the fill dirt. The garage and utility grade elevation will be 103.90' (104.23' less 4" [.33"] = 103.90'). The area will have to be filled to this elevation.

3. The elevation at the right front corner of the garage is 96.96'. A fill of 6.94' [6'11"] must bemade here (103.90' less 96.96' = 6.94').

4. The elevation at the left front corner of the garage is 97.85'. A fill of 6.05' [6'0"] will be needed here (103.90' less 97.85' = 6.05').

5. The elevation at the left back corner of the utility area is 101.40'. A fill

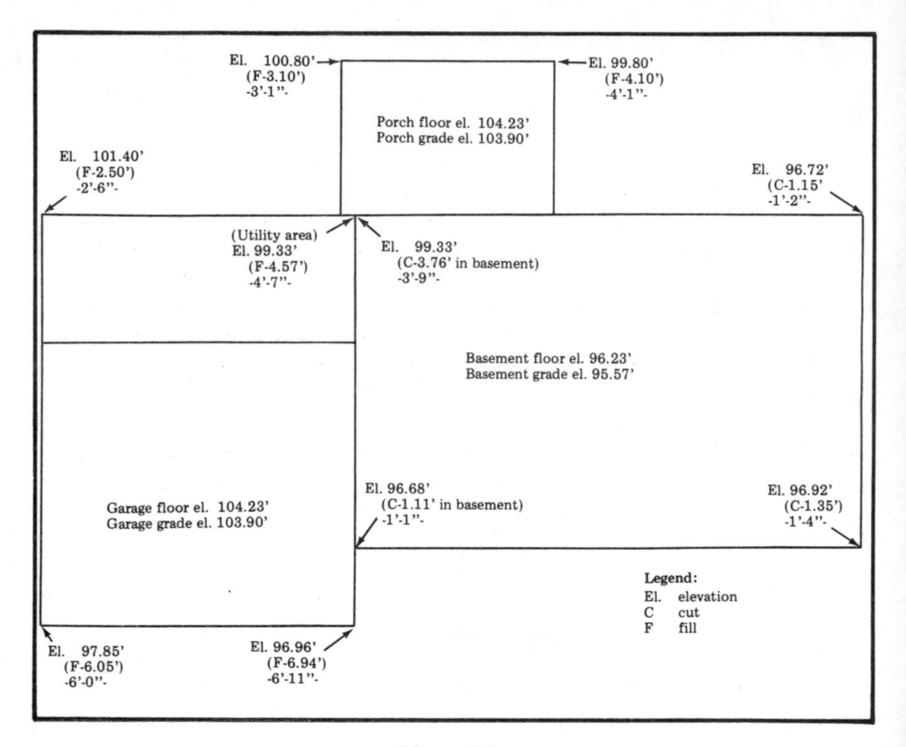

Floor Plan
Figure 3-3

of 2.50' [2'6''] is required here (103.90' less 101.40' = 2.50').

6. The elevation at the right back corner of the utility area is 99.33'. A fill of 4.57' [4'7''] must be made here.

7. The average fill in the garage and utility area is 5.02' [5'0'']. The number of cubic yards of fill dirt that will be required in this area is:

$$\frac{24.33'\ (24'\text{-}4'')\times 32.0'\times 5.02'}{27} = 144.75 \text{ cubic yards}$$

Allowing for compaction (25%)	36.19 cubic yards
Total for garage area	180.94 cubic yards

8. A walk from the front of the garage to the front stoop will be installed. The elevation of this walk and stoop will be the same as the garage and utility floor elevation (104.23'). The finished grade from the left front corner of the garage to the right front corner of the house is to be the same elevation, and it is to taper to the left and right front corners of the lot at the street. The average depth of the fill in the front yard will be 3.61' [3'7'']. The amount of fill dirt that will be required in the front yard will be:

Average width of fill. . . .107.0' [lot 150.0' plus dimension of house 64.0'' = 214.0' divided by 2 = 107.0'].

$$\frac{107.00' \times 77.00' \times 3.61'}{27} = 1101.58 \text{ cubic yards}$$

The offset at the front of the house adjacent to the garage is 39'8'' x 6'0'' and there is an average fill of 7.43' in this area. The fill dirt that will needed here is:

$$\frac{39.67' \, (39'\text{-}8'') \times 6.0' \times 7.43'}{27} = 65.50 \text{ cu. yds}$$

9. The fill dirt for the side and back yards was computed as above and the amount of soil from the excavation plus rough grading in the back yard will take care of these areas.

10. The amount of fill dirt that will have to be imported to the building site will be:

Garage and utility area	180.94 cubic yards
Front yard	1101.58 cubic yards
Front offset	65.50 cubic yards
Side and back yards	-0- cubic yards
	1348.02 cubic yards
(*) Allowfor compaction (25%)	337.01 cubic yards
Total	1685.03 cubic yards

(*) An allowance of 25% for compaction in the garage and utility area has already been made, but is again computed here as a safeguard against underestimating the fill dirt.

Site Cleaning and Hauling

There is some expense in disposing of the waste material after every construction project is completed. Biodegradable material such as scrap lumber, scrap roofing and cement bags may be hauled to an authorized dump site. Toxic or non-biodegradable material such as waste from gypsum boards and empty paint cans should be hauled away from the site and dealt with according to local regulations. Burning waste is banned in most areas. Other items that must be hauled away from the building site at the conclusion of the construction are equipment, scaffolding, and left over material that is reusable.

The expense of cleaning the area and hauling this material away from the site is part of the construction costs and should not be overlooked by the estimator in making his cost estimate. This expense can only be estimated when the cost estimate is made, but past experience and accurate cost records from previous jobs can give the estimator a fairly accurate estimate.

Cost Estimate Work Sheet For
Site Clearing, Excavations and Fill Dirt

3.1	Site clearing	$_______
3.2	Excavation	_______
3.3	Fill dirt	_______
3.4	Site cleaning and hauling	_______
	Total	$_______

(*) Enter on Line 3, Form 100

(*) Form 100 is the Cost Estimate Form shown in the last chapter.

Chapter 4

Footings

The footings support the house as long as it stands, and they must be designed and built for a minimum of settling. Soil conditions vary, but in residential construction a soil bearing analysis is seldom required. The design of the footings on the blueprint and the local building codes are normally sufficient. The minimum width of footing shall be what is required to safely support the total design load without excessive settling. Contributing factors to the design of the footings, other than the size of the house, are the safe bearing value of the soil and its stability. An architect who draws the plans for a particular house for a designated building site will design the footing for that site. Plans furnished with a factory built house or stock plans from a house planning service may show the size of the footings that will support the house under normal soil conditions, but the zoning codes where the house is to be built may require wider and deeper footings due to the local soil conditions. For this reason, many blueprints do not give the dimensions of the footings but leave it to the builder to adapt to the local codes.

Footings must be poured on solid earth. Any footing poured on soft or wet soil will cause cracks in the foundation wall. Footings for garages detached from the main wall, porches or stoops may be poured later. However, the cost estimate should still include the cost.

The construction of the footings affects the cost of the house. If stepped footings are required, the cost will increase. Adverse soil conditions are another factor that will affect the labor output and add to the material costs, and is another reason why square foot and cubic foot estimates are approximate costs only. Conduit, under or through the footing, may be required for the water, sewer, electricity, gas, telephone and television cable utility service lines. The time to install this is during the preparation for pouring the footing. The estimator should not overlook this cost in his estimate.

Construction during the winter increases costs of the footings, a factor the estimator should not overlook. In addition to the extra labor costs, the concrete truck may not be able to get to the site, and a front end loader will have to be used to haul the concrete from the truck to the site. In freezing weather the concrete should be poured as soon as possible after excavation and protected against freezing until hardened. Calcium chloride, a water-absorbing chemical, is normally added to concrete to accelerate the setting time and so reduce the time it is exposed to freezing. Be careful not to add too much calcium chloride, as an excess amount will weaken the concrete. In freezing weather use approximately two pounds of calcium chloride per bag of cement, making adjustments in extreme temperatures.

Layout

Once the excavation has been completed, the foundation must be laid out for the permanent corners before the construction of the footings can begin. Although a rough layout of the foundation was made for the excavation, and reference stakes were probably put in place, it is doubtful if batter boards were constructed (if they are to be used).

The cost of laying out the foundation for the permanent house corners, setting the forms and grade stakes to the correct footing elevation, and leveling them, will vary from job to job. Only if this layout work has been let on a contract can a definite cost be determined in advance. Past cost records can be the only criteria for an accurate estimate.

Estimating Concrete Quantity

Estimating the quantity of concrete required for the footing before the construction begins is easy on paper, but the quantity computed from the blueprints and the quantity that will actually be needed may be two different figures. True, the dimensions of the foundation will not change, but the estimator may neglect to allow for the following:

1. During the excavation for the footing, the excavation frequently will be wider than the specified footing width (the footing should never be less than the specified width). Forms for the concrete can be kept to a true dimension, but when excavation is done by machine or by hand it is difficult to be consistent with the dimensions.

2. Excavation to the exact depth is also difficult. The depth of the concrete should never be less than the specified depth, and if the excavation is more than the required depth it should be filled with concrete and not loose dirt.

3. When stepped footing is required, the amount of concrete used in the vertical rise can be overlooked (Figures 4-2 and 4-3). These three factors are commonly overlooked in estimating concrete for footing, along with allowances made for waste, such as concrete spilling from the truck dur-

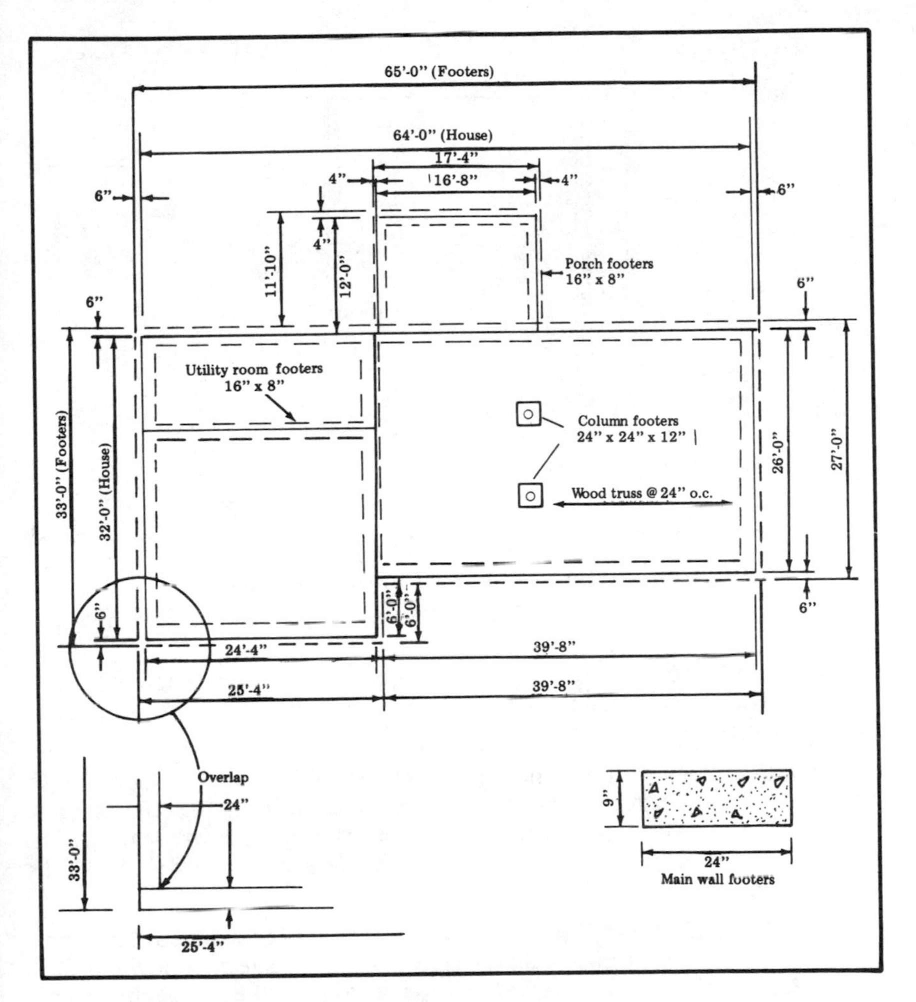

Foundation Footing
Figure 4-1

ing the unloading and from the wheelbarrows. Some of this concrete will be salvaged, but much of it will be wasted. It is often cheaper to pay for a little extra concrete in the beginning than pay the extra cost for a small amount that may be needed to complete the job. Most ready-mix concrete companies have a minimum charge for delivery and the cost of delivering less than the minimum amount can be high. Less than one cubic yard of concrete needed to complete a job will require a special trip to the job by the supplier, and this small amount of concrete can cost as much as three or four cubic yards delivered during the regular delivery. Many estimators double the corners, or do not allow for the overlap, in estimating concrete quantities to help offset for the waste (Figure 4-1).

Before estimating the concrete, make an examination of the grade

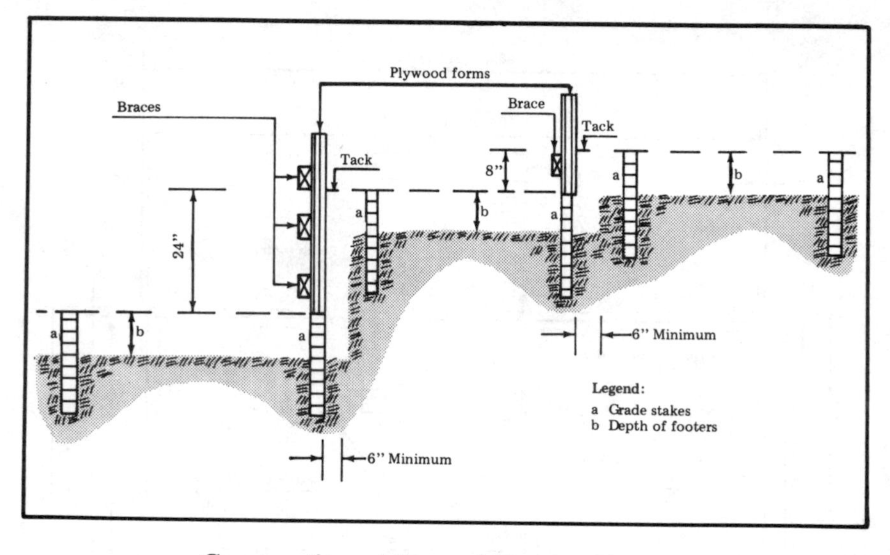

Construction of Stepped Footing Forms
Figure 4-2

elevations (Figure 3-1) to determine if stepped footings will be required. If so, a close estimate can be made for the concrete that will be needed in the vertical rise as follows:

1. From Figure 3-1, the basement grade elevation is shown as 95.57' and the grade elevation at the left back corner of the utility area (the garage and utility area will require fill dirt and the footing here will start from the present grade elevation) is 101.40'. This is a rise of 5.83' [5'-10"] (101.40' less 95.57' = 5.83').

2. Several steps in multiples of 8" will have to be constructed from the left front corner of the basement (adjacent to the garage area) around the perimeter of the front and left side of the garage, to the left back corner of the utility area. The sum of the vertical rise for these steps (in multiples of 8") should be 72" [5'-10" is 70" and is not a multiple of 8"]. The required amount of concrete for the vertical rise in the stepped footing will be 24" (width of footings) x 6" (vertical thickness of footings) x 72" (vertical rise) x 2 (front and back perimeter) divided by 27, as shown below:

$$\frac{2.0' \,(24") \times .50' \,(6") \times 6.0' \,(72") \times 2}{27} = .44 \text{ cubic yards}$$

This is not a large amount of concrete, but if the concrete truck has to make a special trip to the jobsite to deliver it because it was not estimated in the beginning, it can be very costly.

From Figures 4-1, 4-2 and 4-3, the amount of concrete can now be estimated. The dimensions of the footings, except in the areas shown, are

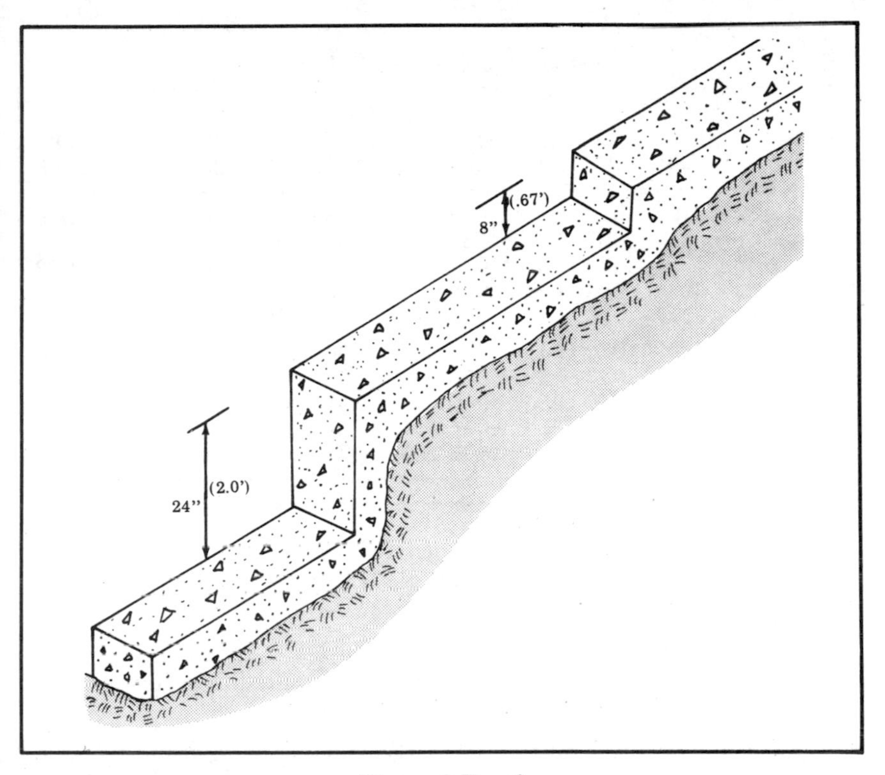

Stepped Footing
Figure 4-3

24" wide and 9" in depth. The total linear feet of the main footings (disregarding the overlaps) as shown in Figure 4-1 is:

Left side	33'-0"
Back	65'-0"
Right side (27'-0" + 6'-0")	33'-0"
Front	65'-0"
Total lin. ft.	196'-0"

$$\frac{196.0' \text{ x } 2.0' \text{ (24") x } .75' \text{ (9")}}{27} = 10.89 \text{ cubic yards}$$

The footings for the back porch and utility area are 16" x 8" and the concrete that will be required for them will be:

Back porch (11'-10" + 17'-4" + 11'-10"	41'-0"
Utility area	24'-4"
Total lin. ft.	65'-4"

$$\frac{65.33' \text{ (65'-4") x } 1.33' \text{ (16") x } .67' \text{ (8")}}{27} = 2.16 \text{ cubic yards}$$

Factors For Concrete Volume											
		Footing Width									
Footing Depth		12" 1.00'	14" 1.17'	16" 1.33'	18" 1.50'	20" 1.67'	22" 1.83'	24" 2.00'	26" 2.17'	28" 2.33'	30" 2.50'
6"	.50'	.019	.022	.025	.028	.031	.034	.037	.040	.043	.046
7"	.58'	.021	.025	.029	.032	.036	.039	.043	.047	.050	.054
8"	.67'	.025	.029	.033	.037	.041	.045	.050	.054	.058	.062
9"	.75'	.028	.033	.037	.042	.046	.051	.056	.060	.065	.069
10"	.83'	.031	.036	.041	.046	.051	.056	.061	.067	.072	.077
11"	.92'	.034	.040	.045	.051	.057	.062	.068	.074	.079	.085
12"	1.00'	.037	.043	..049	.056	.062	.068	.074	.080	.086	.093
13"	1.08'	.040	.047	.053	.060	.067	.073	.080	.087	.093	.100
14"	1.17'	.043	.051	.058	.065	.072	.079	.087	.094	.101	.108
15"	1.25'	.046	.054	.062	.609	.077	.085	.093	.100	.108	.116

Factors for Concrete Volume
Figure 4-4

The blueprints show two column footings, each with dimensions of 24" x 24" x 12".

$$\frac{2.0'\,(24'')\times 2.0'\,(24'')\times 1.0'\,(12'')\times 2}{27} = .30 \text{ cubic yards}$$

The total amount of concrete required for the footing will be:

Main wall	10.89 cubic yards
Porch and utility area	2.16 cubic yards
Vertical rise in stepped footers	.44 cubic yards
Column footers	.30 cubic yards
Total	13.79 cubic yards

Figure 4-4 is a chart showing a table of factors that may be used in estimating concrete quantities. By multiplying the factor shown in the footing width and depth column by the total linear feet of the footings, the number of cubic yards of concrete that will be required can be readily estimated.

Example: Find the number of cubic yards of concrete that will be required for a footing that is 16" x 8" and is 65'-4" in length. In Figure 4-4, go across the footing width line to the 16" column. In the 16" column, go down to the footing depth line to 8". Where these two lines intersect the number .033 will be found. Multiply this factor by the total linear feet to get the number of cubic yards of concrete that will be required.

.033 x 65.33' (65'-4") = 2.16 cubic yards

Check:
$$\frac{65.33'\,(65'\text{-}4'')\times 1.33'\,(16'')\times .67'\,(8'')}{27} = 2.16 \text{ cubic yards}$$

The number of cubic yards of concrete required for column, or pier, footing can be computed as follows:

Example: Column footings 24" x 24" x 12" are 24" wide, 12" in depth and 24" in length. The factor for 24" x 12" is .074.

.074 x 2.0' (24" in length) = .15 cubic yards

If there are 5 column footings the total cubic yards of concrete required would be:

.074 x 2.0' x 5 = .74 cubic yards

Check: $$\frac{2.0'\ (24'')\ \text{x}\ 2.0'\ (24''\ \text{x}\ 1.0'\ (12'')\ \text{x}\ 5}{27} = .74\ \text{cubic yards}$$

If the footings are wider than the width shown in the tables in Figure 4-4, as would be the case for chimney footings, the factors can be added together to make up this width.

Example: A chimney footing is to be 3'0" (36") wide, 6'0" in length and 12" (1.0') in depth.

The factor for 20" x 12" is .062
The factor for 16" x 12" is .049
The factor for 36" x 12" is .111

.111 x 6.0' = .67 cubic yards

Check: $$\frac{3.0'\ (36'')\ \text{x}\ 6.0'\ \text{x}\ 1.0'\ (12'')}{27} = .67\ \text{cubic yards}$$

Note: In computing the cubic yards where large quantities of concrete will be required, there may be a slight difference (using the factors in Figure 4-4) from the mathematical formula shown in the check above. This mathematical formula is the prescribed method for computing cubic yards; therefore, these tables should be used only for a quick reference when computing large quantities of concrete.

Estimating Other Material

The size and number of reinforcing rods to be placed in the footing will be shown on the wall section of the blueprints (Figure 4-5), but the local zoning regulations may require that the size and number be increased because of soil conditions at the building site. The greater of the two requirements must be used.

The size of the reinforcing rods will be referred to by their diameter in inches, as ½" diameter rods, or by their numerals, as #4 diameter rods. The bar numbers are based on the number of one-eighth inches (1/8") in the diameter of the rod. Thus, #4 diameter rods are 4/8" or ½" in diameter; and #3 diameter rods are 3/8" in diameter (Figure 4-6).

After computing the total linear feet of reinforcing rods that will be required for the footing, add 5% to allow for overlapping. Divide the total linear feet by 20 to get the number of rods required.

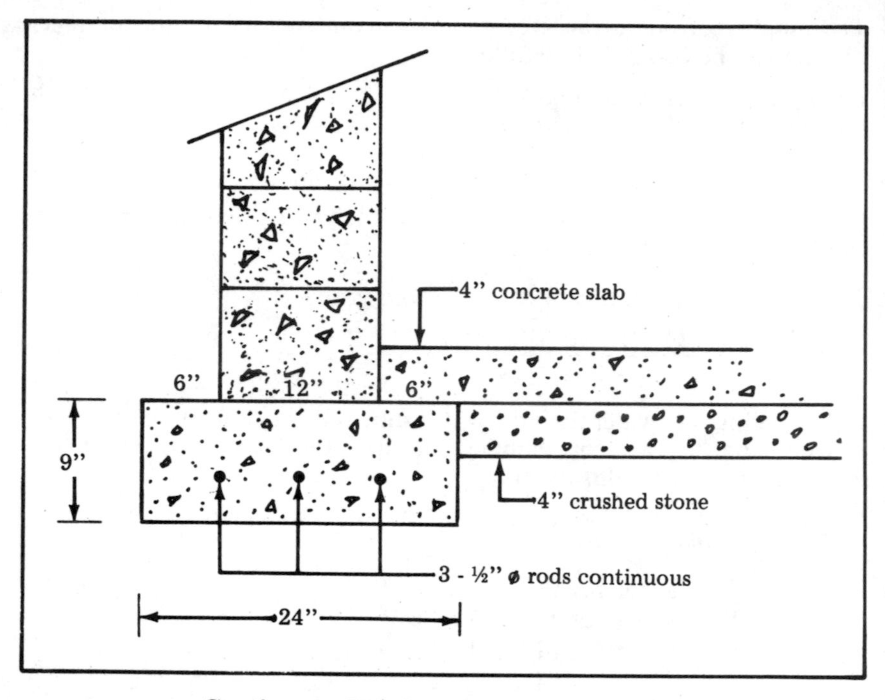

Continuous Reinforcing Rods in Footing
Figure 4-5

Reinforcing Rods

Rod Size		Weight
Diameter Inches	Rod Number	Pounds Per Foot
1/4"	2	.167
3/8"	3	.376
1/2"	4	.668
5/8"	5	1.043
3/4"	6	1.502
7/8"	7	2.044
1"	8	2.670

Note: If the reinforcing rods are sold by weight, compute the total weight and cost as follows: 1. Multiply the total number of linear feet by the weight per foot. ***Example:*** The weight of 940 linear feet of ½" diameter rods is 627.92 pounds (940 x .668 = 627.92 lbs.). 2. Total weight (rounded off) multiplied by the rate = cost.

Table of Rod Numbers and Weights
Figure 4-6

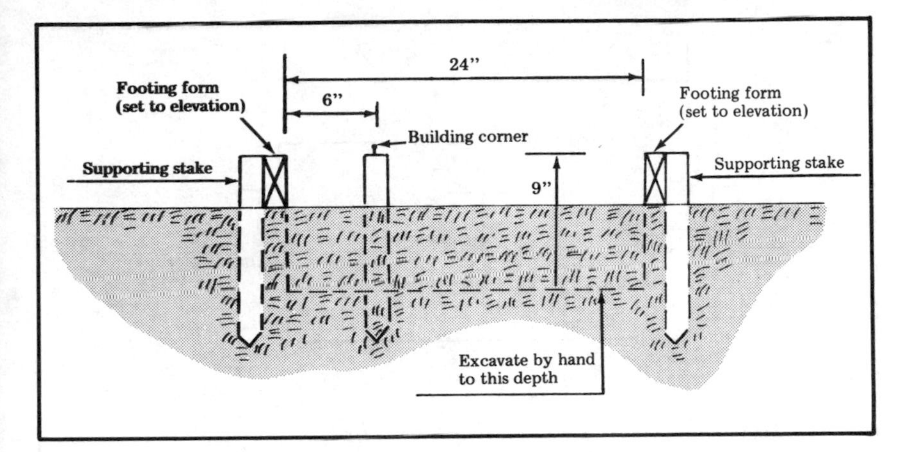

Forms for Footing Before Excavation
Figure 4-7

Example: The total linear feet of the footing from the blueprints and shown in Figure 4-1 is:

Perimeter of Foundation

Left side	33'-0"
Back	65'-0"
(27'-0" + 6'-0") =	
Right side	33'-0"
Front	65'-0"
Utility room footing	24'-4"
(11-10" + 17'-4" + 11'-10")	
Back porch	41'-0"
(2 x 24")	
Column footing	4'-0"
(2 x 72")	
Vertical rise in steps	12'-0"
	277'-4"

The number of continuous rods as shown in Figure 4-5 is 3. The total linear feet of reinforcing rods will be:

3 x 277.33' (277'-4")	=	831.99'
Grade stakes (30 stakes @ 20")	=	50.00'
(See Figure 4-10)		881.99'
Allow for overlapping (5%)	=	44.10
		926.09'

926.09' ÷ 20' = 46.30 or 47 pieces

Order: 47 pieces ½" (No. 4) diameter rods or 940 lin. ft.

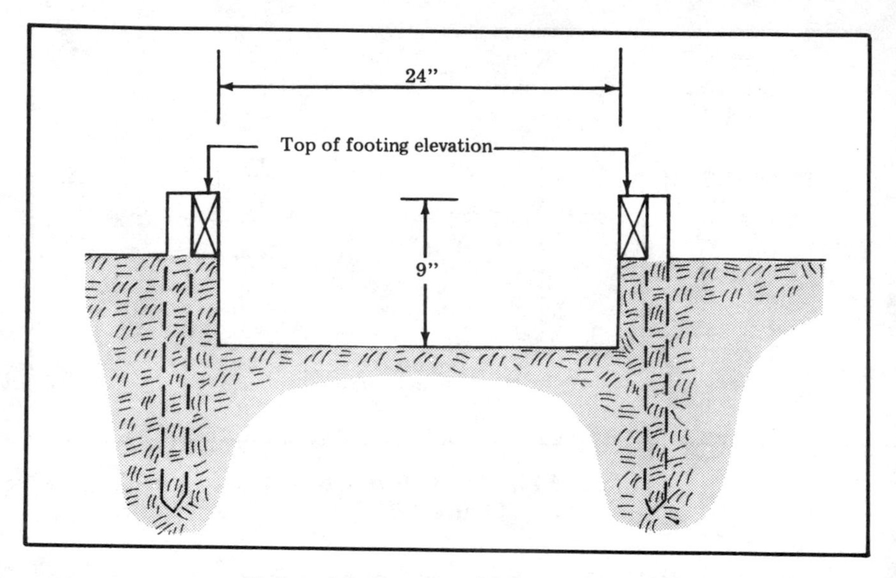

Forms for Footing After Excavation
Figure 4-8

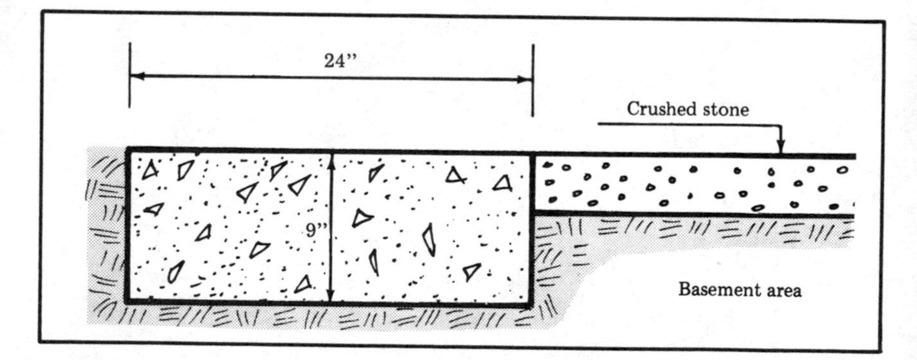

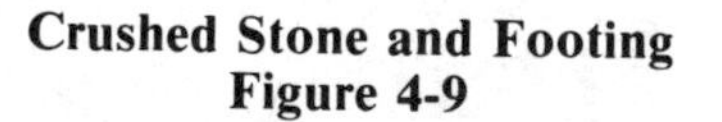

Crushed Stone and Footing
Figure 4-9

Forms will be constructed for the footing in the basement area as shown in Figure 4-7 and 4-8. Using 2" x 4" framing and forming the footing as shown, the stone under the concrete floor will be level with the top of the footings as shown in Figures 4-5 and 4-9.

The perimeter of the basement footing from the plans and Figure 4-1 is:

Front	39'-8"
Left side	27'-0"
Back	39'-8"
Right side	27'-0"
Column footers (2 x 24" x 24" each)	16'-0"
	148'-16" = (149'-4")

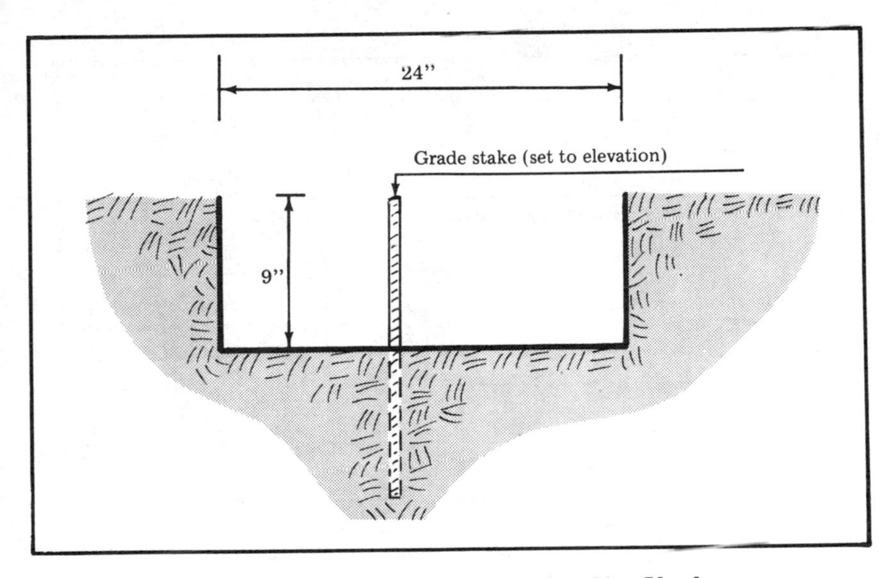

Grade Stakes When Forms Are Not Used
Figure 4-10

149.33' (149'-4") ÷ 12 = 12.44 (or 13 - 2" x 4" x 12')

Adding the supporting stakes for the forms, and the braces for the vertical rise in the stepped footings around the garage and utility areas, an additional 12 - 2" x 4" x 12' will be required, making a total of 25 - 2" x 4" x 12' for the footing.

The plywood that will be required for the stepped footing, (Figure 4-2), will be 2 pieces, ½" x 4' x 8' C-D. An allowance can be made for the wire to tie the reinforcing rods together and the nails for the forms.

When the forms are constructed for the footing, the top of the forms should be the calculated footing elevation. When there are no forms (Figure 4-10), grade stakes cut about 20" in length from reinforcing rods should be used. The top of these grade stakes should be set to the footing elevation.

Note: Much of the material used for the forms can be salvaged and is reuseable. The cost of this material may be prorated over two or more jobs.

Estimating Labor Costs

Material can be estimated with a high degree of accuracy. ***Example:*** 8" x 16" concrete blocks require 3 blocks per course for 4 linear feet. The number of blocks per course = linear feet x 75%; 60 linear feet will require 45 blocks per course (60 x 75% = 45 blocks). However, labor can only be estimated, and this is usually where the contractor makes or loses money on a house. Labor productivity varies with each different craftsman and with each different day. There are many guidelines published for estimating labor costs on each phase of con-

For Period Ending 8-21-76 Brown Job

Line #	NAME	Exemptions	Days August 16 M	17 T	18 W	19 T	20 F	21 S	Rate	Hours Worked Reg.	Over Time	Total Earnings
1	D. White		X	X	3	8	8	X		19		
2	R. Kidd		X	X	3	8	8	X		19		
3	R. Farlow		X	X	3	8	8	X		19		
4												
5												
6												
7												
8												
9												
10												
11												
12												
13												
14												
15												
16												
17												
18												
19												
20												

Daily Log

Monday —

Tuesday —

Wednesday Laid out Footers.

Thursday Forms for Footers in basement area.

Friday Forms for Footers in basement area.

Saturday —

Weekly Time Sheet
Figure 4-11

structing a house, but they are only guidelines. The most accurate method of estimating labor costs is to keep a daily log on the weekly payroll sheet on each job, (Figure 4-11), and use it in estimating future labor costs on similar jobs. Allowances must be made in any cost

For period ending 8-28-76   Brown Job

Line #	Name	Exemptions	Days (August) 23 M	24 T	25 W	26 T	27 F	28 S	Rate	Hours Worked Reg.	Over Time	Total Earnings
1	D. White		8	8	6	6½	2½	X		31		
2	R. Kidd		8	8	6	6½	2½	X		31		
3	R. Farlow		8	8	6	6½	X	X		28½		
4	N. Neel		X	X	X	6½	X	X		6½		
5												
6												
7												
8												
9												
10												
11												
12												
13												
14												
15												
16												
17												
18												
19												
20												

Daily Log

Monday Forms for stepped footers in garage area.

Tuesday Forms for stepped footers in garage area.

Wednesday Finished stepped footers — Ready for concrete.

Thursday Poured concrete for footers (14 cu. yds.).

Friday Stripped forms from footing.

Saturday —.

Weekly Time Sheet
Figure 4-12

estimate for labor in abnormal conditions. Winter weather can increase labor costs for footing from 10 to 15%. Here again, accurate cost records from previous jobs are essential.

A typical example of why accurate cost records from previous jobs are essential is as follows:

1. A backhoe should excavate approximately 75 - 100 cubic yards per day under ordinary conditions.

2. Cost records from a recent job reveal that the backhoe on that job using a 24" bucket took 5½ hours to excavate 6.4 cubic yards of earth for the stepped footing. Adding 1 hour for the travel time to the building site means it took 6½ hours to excavate 6.4 cubic yards of earth, or 1 cubic yard per hour. Knowing this can save many dollars lost from underestimating labor and machine costs.

To compute the labor costs to form, hand excavate, pour concrete and strip the forms from the footing shown in Figure 4-1 proceed as follows:

1. Refer to the time sheets of a similar house constructed recently as shown in Figure 4-11 and 4-12.

2. The daily log at the bottom of these time sheets reveals work was started on the forms on Thursday, August 19th and completed on Wednesday, August 25th. The concrete was poured on Thursday, August 26th (an extra man was used this day to help pour the concrete because some of it has to be placed with wheelbarrows). On Friday, August 27th two men stripped the forms from the footing. A total of 145 manhours beginning Thursday, August 19th thru Friday, August 27th was required to form, hand excavate, pour concrete and strip the forms for 12.91 cubic yards (rounded off to 13), including the stepped footing.

3. Use the following formula to compute the future labor costs:

$$\frac{\text{Total man-hours}}{\text{Cubic yards}} = \text{man-hours}$$

$$\frac{\text{145 man-hours}}{\text{13 cubic yards}} = \text{11.15 man-hour factor}$$

4. The cubic yards computed for the footing in Figure 4-1 is 13.79 (rounded off to 14). The estimated number of man-hours that will be required to form, hand excavate, pour and strip the forms will be:

Man-hour factor x cubic yards = Total man-hours
11.15 (Factor) x 14 (cu. yds.) = 156.10 man-hours

5. Three men will be assigned to this job and the allotted time for each man will be:

$$\frac{\text{Total man-hours}}{\text{Number of men}} = \text{time allotted to each man}$$

$$\frac{\text{156.10 (man-hours)}}{\text{3 (number of men)}} = \begin{array}{l}\text{52.03 (rounded off to 52)}\\ \text{hours allotted to each}\\ \text{man for the footing}\end{array}$$

6. The wages for the foreman will be $8.50 per hour, and the wages for the two helpers will be $6.25 per hour each. The labor costs can be estimated as follows:

Foreman	52 man-hours	@ $8.50	=	$442.00
Helper	52 man-hours	@ 6.25	=	325.00
Helper	52 man-hours	@ 6.25	=	325.00
	156 man-hours	@	=	$1092.00 (*)

(*) Enter this amount on Line (1) on the cost estimate work sheet for footers. ***Note:*** The F.I.C.A., F.U.T.A., Workman's Compensation and Liability insurance costs will be added later when the total labor costs are known.

The cost of the machine excavation from previous cost records is $180.00, and this estimate is entered on Line (k) on the cost estimate worksheet.

Cost Estimate Work Sheet for Footers

4.1 Layout $ _____

4.2 (*Concrete Quantity:*

(a) Size: _____ width; x _____ depth; x _____ lin. ft. = _____ cu. ft.

Size: _____ width; x _____ depth; x _____ lin. ft. = _____ cu. ft.

Column Footers: Number _____.
Size: _____ w; x _____ d; x _____ lin. ft. x _____ no. = _____ cu. ft.

Chimney Footers: Number _____.
Size: _____ w; x _____ d; x _____ lin. ft. x _____ no. = _____ cu. ft.

Vertical Rise For Stepped Footers:
Size: _____ w; x _____ thickness; x _____ height = _____ cu. ft.

Other (Specify) _____ cu. ft.

Total _____ cu. ft.

Total cubic feet _____ ÷ 27 = _____ cubic yards

(b) Order (rounded off) _____ cubic yards

(c) Cost of concrete (test: _____ psi) $_____

4.3 *Other Material:*

*(d) Linear feet of reinforcing rods _____

(e) Cost of reinforcing rods $_____
(Size: _____)

(f) Number of stepped footers _____

(g) Average height of steps _____

(h) Cost of plywood forms $_____
(Quantity and size: _____

(i) Cost of framing lumber $_____
(Quantity and size: _____)

(j) Cost of other material $_____
(Specify: _____)

4.4 *Labor Costs:*

(k) Estimated cost of excavation $_____

(l) Estimated cost to form and pour $_____

(*) *If grade stakes are cut from the reinforcing rods 20" in length allow 12 stakes for each 20' piece of reinforcing rod.*

Brought forward from page 1 $________

(b) ____ cu. yds. x (c) $____ = $________
(Cost of concrete)
(d) ____ lin. ft. x (e) $____ = $________
(Cost of reinforcing rods)

Add line (h) $________
Add line (i) $________
Add line (j) $________
Subtotal $________
Sales tax (____%) $________
Cost of material $________ $________
Add line (k) ________
Add line (l) ________
Cost of footers $________
* *(Enter on Line 4, Form 100)*

* *Cost Estimate Form*

Chapter 5

Foundations

When there is a basement and the plans specify crushed stone under the concrete basement floor, it is much less expensive to haul the stone to the basement area and spread it before the foundation masonry blocks are started. The trucks hauling the stone can get as close to the foundation area as the concrete trucks that delivered the concrete for the footing. It then becomes an easy task to spread the stone. If this crushed stone is not placed before the foundation walls are constructed it will have to be hauled to the area in wheelbarrows. In addition to saving money by placing this stone before the masonry blocks are started, it will also provide the masons and their helpers a safer and more satisfactory area to work in.

Crushed stone is usually sold by weight, and priced by the ton. Occasionally it is sold by volume and priced by the cubic yard. To compute the volume and weight of the crushed stone required for the basement area in Figure 4-1, proceed as follows:

1. The outside dimensions of the basement area are 39'-8" x 26'-0".

2. The inside dimensions are 38'-8" x 24'-0" (39'-8" less the thickness of the block wall on the right side [12"], and 26'-0" less the thickness of the block wall in the front and back [12" + 12"].

3. The footing was formed in the basement area for the crushed stone to be level with the top of the footings (Figures 5-1 and 5-2). The plans specify 4" of stone (Figure 5-2).

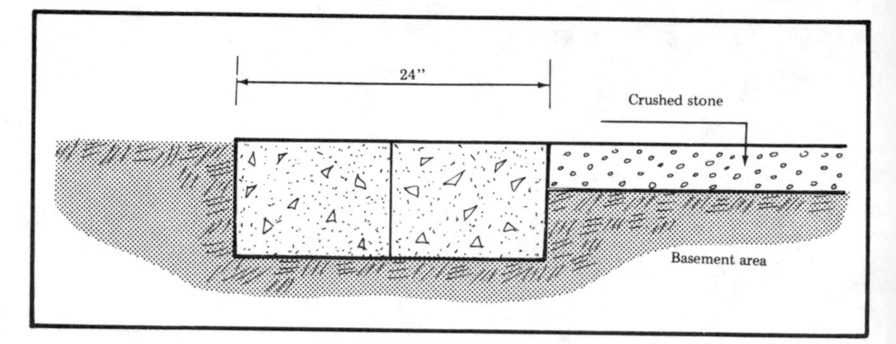

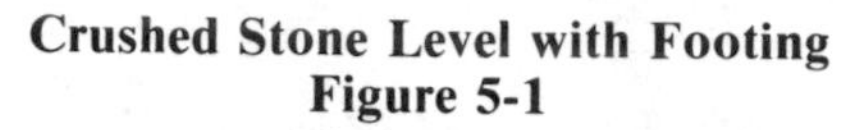

Crushed Stone Level with Footing
Figure 5-1

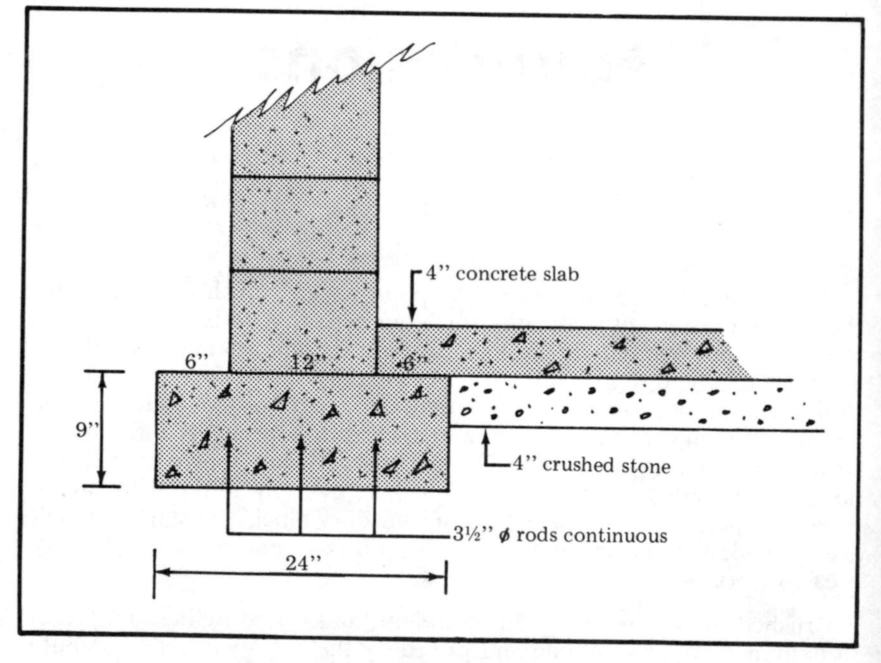

Concrete Slab Over Crushed Stone
Figure 5-2

4. The cubic yards of stone for the basement area will be:

$$\frac{38.67' \ (38'\text{-}8'') \times 24.0' \times .33' \ (4'')}{27} = 11.34 \text{ cubic yards}$$

5. The weight of crushed stone is 2700 lbs. = 1 cu. yd.

$$\frac{11.34 \text{ cubic yards x 2700 pounds}}{2000 \text{ (pounds per ton)}} = 15.31 \text{ tons (*)}$$

(*) An allowance should be made for extra stone because of normal uneven grade excavation.

Estimate (with allowance) 18 tons crushed stone

6. The cost of the stone delivered to the job site has been quoted by the supplier at $5.30 per ton.

18 tons @ $5.30 = $95.40

(Enter on Line 5.1 on Cost Estimate Work Sheet)

The labor to spread the stone is computed as follows:

1. The daily log from the payroll sheets of previous jobs show that an average of .65 man-hours is required to spread one ton of crushed stone.

2. 18 tons crushed stone x .65 = 11.7 or 12 man-hours.

12 manhours x rate = labor cost

(Enter on Line 5.9 Cost Estimate Work Sheet)

Estimating Masonry Blocks

The estimator should always have the foundation plan, specifications and the plot plan showing the grade elevations, before he estimates the masonry blocks that will be required for the house foundation. The foundation plan shows the dimensions of the foundation and if there is to be a basement, crawl space or both. It will also give the height of the foundation walls, the size and location of the windows and doors and the number and size of the piers and pilasters. The specifications will specify if the brick will be below grade. From this informaton the estimator will know on what course of blocks the brick will start, and so how many courses of 12" x 8" x 16" masonry blocks to estimate. The plot plan showing the grade elevations will show the estimator how to calculate the stepped footings, if one is necessary (Figure 2-1).

Estimating masonry blocks from the same footing elevation when the number of block courses do not vary is not difficult, but when there are stepped footings (Figure 5-9) an accurate estimate becomes more difficult.

The actual size of masonry blocks is reduced from the nominal size so the block plus one mortar joint (usually 3/8") will equal the nominal size.

Nominal Size	Actual Size
4" x 8" x 16"	3-5/8" x 7-5/8" x 15-5/8"
8" x 8" x 16"	7-5/8" x 7-5/8" x 15-5/8"
12" x 8" x 16"	11-5/8" x 7-5/8" x 15-5/8"

The actual size of a 8" x 8" x 16" masonry block is shown below:

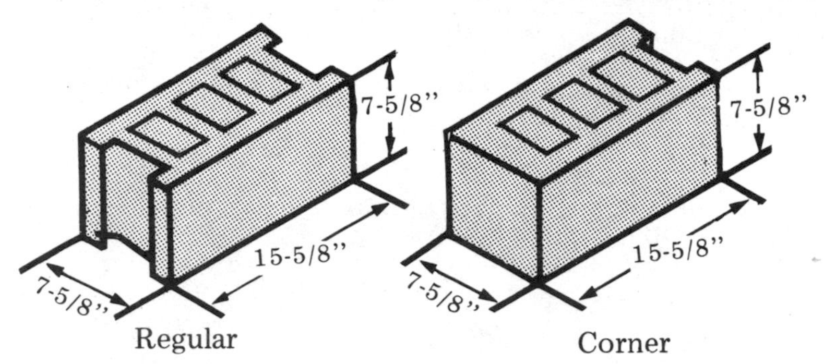

1. One regular or corner block plus one 3/8" mortar joint equals 16" in length.

2. One regular or corner block plus one 3/8" mortar joint equals 8" in height.

The number of blocks required per course is calculated by muitiplying the length of the wall in feet and inches by .75 (75%). *Example:*

1. A wall 4'0" in length will require 3 blocks per course (4.0 x .75 = 3).

2. A wall 39'8" in length will require 29.75 blocks per course. (39.67' [39'8"] x .75 = 29.75).

The total length of the foundation (for each block size) multiplied by .75, multiplied by the number of courses will give the number of blocks for the foundation. *Example:*

1. The basement area of a house foundation is 39'-8" x 26'-0". The plans show 12 courses of regular blocks plus 1 course of 4" x 8" x 16" solid blocks. The brick will start below grade and calculations made from the grade elevations determine that the brick will start on top of the 8th course of blocks. There will then be 8 courses of 12" x 8" x 16" blocks, 4 courses of 8" x 8" x 16" blocks and 1 course of 4" x 8" x 16" solid blocks.

2. The total linear feet of the perimeter of the basement foundation wall will be 131'-4" (39'-8" + 26'-0" x 2 = 131'-4").

3. The number of masonry blocks that will be required will be:

$$\frac{12\text{" x 8" x 16"}}{131.33\text{' (131'-4") x .75 x 8 courses}} = 788 \text{ blocks}$$

$$\frac{8\text{" x 8" x 16"}}{131.33\text{' (131'-4") x .75 x 4 courses}} = 394 \text{ blocks}$$

$$\frac{4\text{" x 8" x 16"}}{131.33\text{" (131'-4") x .75 x 1 course}} = 98.5 \text{ blocks (or 99)}$$

Note: No allowance is made for the overlapping at the corners. These extra blocks will be used to apply against the waste.

Another method used by many estimators and builders in estimating blocks is by the square feet. It requires 112½'' blocks (8'' in height x 16'' in length) per 100 square feet of wall space. *Example:*

1. A wall 25'-0'' in length and 4'-0'' in height is 100 square feet. The blocks required for this wall are: 25.0' x .75 x 6 courses (4'-0'' high) = 112½''. 112½'' divided by 100 = 1.125 blocks per square feet (factor).

Comparing the number of blocks estimated for the same foundation above using the factor for the square feet method reveals the following:

131.33' (131'-4'') x 5.33' (8 courses) = 700 square feet.
700 square feet x 1.125 (factor) = 787.5 *(or 788)* 12'' x 8'' x 16'' C.B.

131.33' (131'-4'') x 2.67' (4 courses) = 350.65 square feet
350.65 square feet x 1.125 (factor) = 394.48 *(or 395)* 8'' x 8'' X 16'' C.B.

131.33' (131'-4'') x .75 x 1 = 98.5 *(or 99)* 4'' x 8'' x 16'' solid C.B.

Viewing the comparison above, either method of estimating the blocks, when the number of block courses in the foundation height does not vary, is acceptable. However, stepped footing is necessary in many foundations (Figure 5-9), and estimating masonry blocks for them is more difficult. Many estimators guess at the number of blocks displaced by the stepped footing. *Example:* Allowing a 50% deduction (or addition), a 25% deduction, etc. This is only a guess, it may be close and it may not even be in the "ball park." A more accurate way to estimate masonry blocks for stepped footing is shown in Figure 5-3, 5-4 (the scale drawing of a masonry block wall), 5-5, 5-6, 5-7 and 5-8. *Example:*

1. The garage floor elevation in Figure 2-1 (the plot plan) is 104.23', indicating a fill will be required in this area.

2. In Figure 5-3, a layout of the stepped footing is sketched.

3. The 6'-0'' offset from the basement wall to the right front corner of the garage shows a rise of 1.06' (13'' rounded off). This was computed by subtracting the baseline footing elevation in the basement (95.90') from the grade elevation at the right front corner of the garage (96.96') [96.96' less 95.90' = 1.06']. 1.06' = 13'' (rounded off). 1-8'' stepped footing will be constructed here 3'-0'' from the garage corner toward the basement wall.

4. The grade elevation rises .89' (97.85' less 96.96' = .89' [11'' rounded off]) from the right front corner to the left front corner of the garage. 1-16'' stepped footing will be constructed here approximately at the halfway mark.

5. From the left front corner of the garage to the left back corner of the utility area, the grade elevation rises 3.55' (101.40' less 97.85' = 3.55'). This is 3'-7'' rounded off, or 43''. Four stepped footings will be constructed for this left side wall. The first stepped footing will be 16'' high,

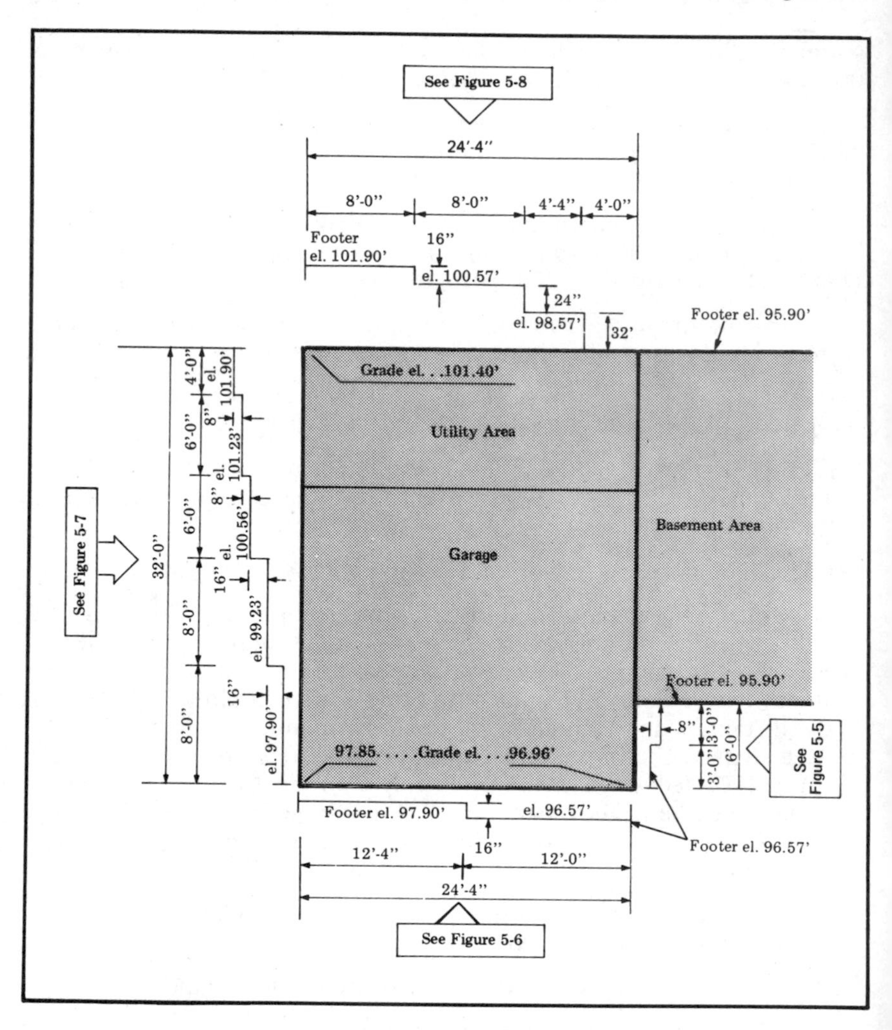

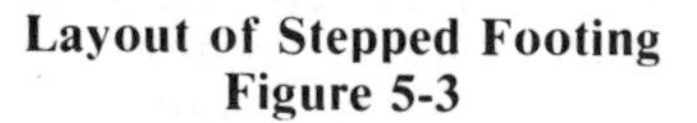

Layout of Stepped Footing
Figure 5-3

approximately 8'-0" from the left front corner, the second step will also be 16" high, 8'-0" from the first step. The third and fourth steps will be 8" in height each; the third step 6'-0" from the second step, and the fourth step will be 6'-0" from the third step and 4'-0" from the left back corner. This will be a total rise for the footings on the left side of 4'-0" (footing el. 101.90' less 97.90' = 4.0' [or 48"]). The top of the footings at the left back corner of the foundation will be .50' (6") higher than the present grade (101.90' footing el., less 101.40' grade el. = .50'). This will be no problem because the finish grade elevation will be above the footing.

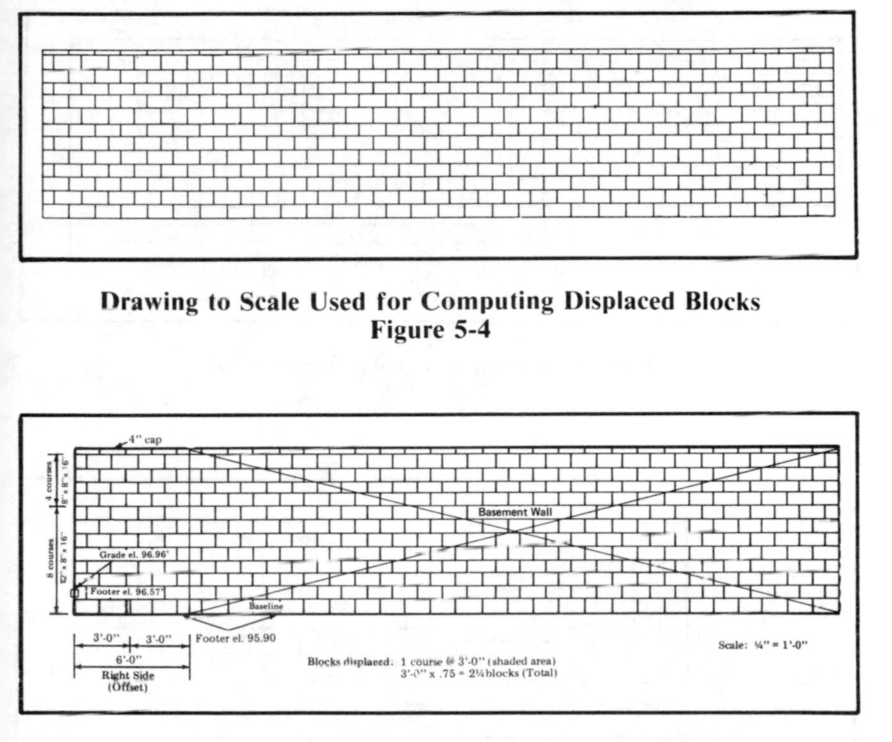

Drawing to Scale Used for Computing Displaced Blocks
Figure 5-4

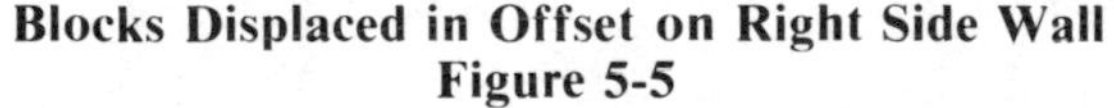

Blocks Displaced in Offset on Right Side Wall
Figure 5-5

6. From the left back corner of the utility area to the right back corner adjacent to the basement area there is a fall of 6.0' (6'-0" or 72") in elevation. This was computed as follows: the footing elevation at the left back corner of the utility area is 101.90' and the basement footing el. is 95.90' (101.90' less 95.90' = 6.0' [or 72"]. This is a fall of 72" in 24'-4" so fewer and higher footing steps will be required. The first footing step will be 16" in height, approximately 8'-0" from the left back corner; the second step will be 24" in height, 8'-0" from the first step, and the third step will be 32" high, 4'-4" from the second step, and 4'-0" from the basement wall.

Figure 5-4 is a masonry block wall drawn to scale. Reproducing a similar drawing (drawn to scale) can be very helpful in estimating the masonry blocks for the walls where stepped footing will be required. *Example:*

1. Figure 5-3 shows 1-8" stepped footing in the 6'-0" offset from the basement wall to the right front corner of the garage. Figure 5-5 is a scale drawing showing this stepped footing (shaded area). The calculations on the drawing show a total of 2¼ blocks will be deducted from this wall.

2. From Figure 5-3, one 16" stepped footing will be required across the front of the garage. Figure 5-6 is a scale drawing of this footing (shaded

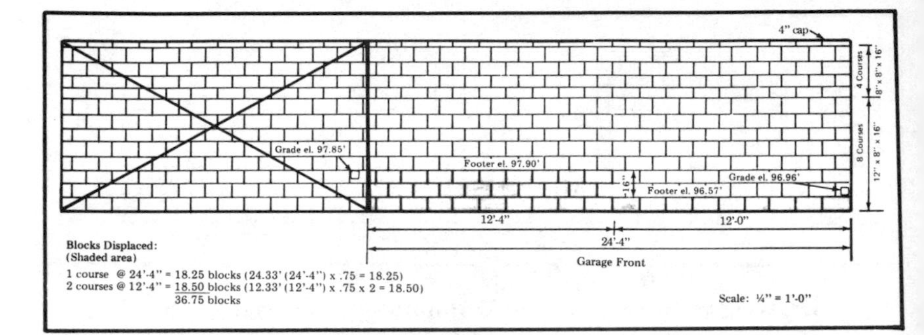

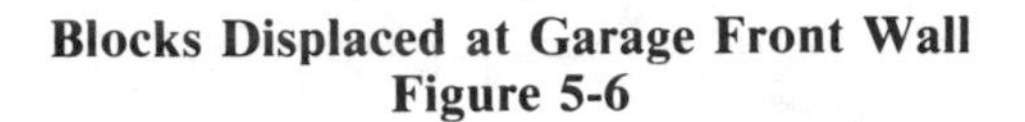

Blocks Displaced at Garage Front Wall
Figure 5-6

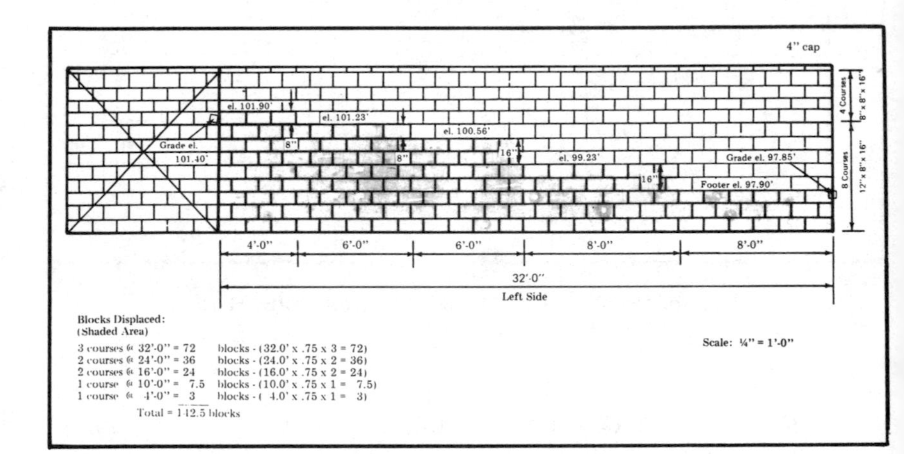

Blocks Displaced on Left Wall
Figure 5-7

area). The calculations on Figure 5-6 show a total of 36.75 blocks will be displaced from this wall by the stepped footing.

3. From Figure 5-3, four stepped footings will be required along the left side. Figure 5-7 is the scale drawing of this wall. The calculations on the drawing show 142½" blocks will be displaced by the stepped footing (shaded area) from this wall.

4. Again referring to Figure 5-3, three stepped footings will be required along the back wall of the utility area down to the basement wall. Figure 5-8 is a scale drawing of this wall. The calculations on this drawing show 109 blocks will be displaced from the back wall.

The masonry blocks that will be needed for the garage and utility area (Figures 5-3, 5-5, 5-6, 5-7 and 5-8), can now be estimated with reasonable accuracy:

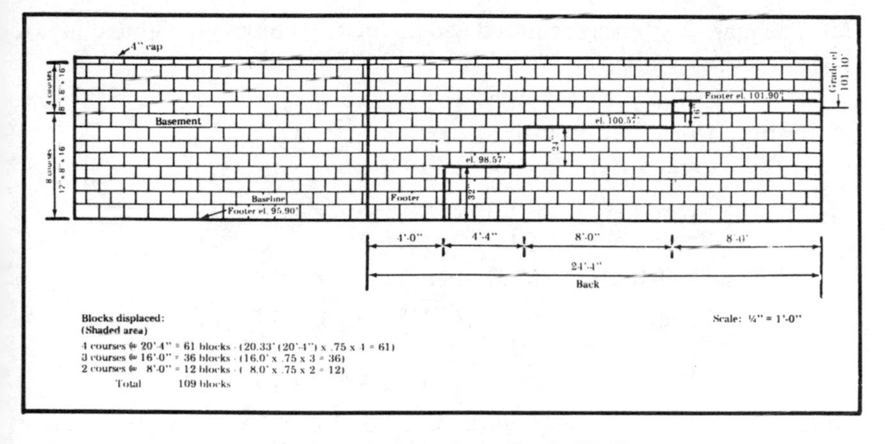

Blocks Displaced on Back Wall
Figure 5-8

1. Compute the quantity of blocks that would be required if there were no stepped footing.

2. From the scale drawing in Figures 5-5, 5-6, 5-7 and 5-8 showing the displaced blocks from the stepped footing (shaded areas), deduct these blocks from the total blocks as computed in 1 above. *Example:*

(a) The linear feet of masonry walls in the garage and utility area requiring stepped footing as shown in Figure 5-3 is 86'-8" (6'-0" + 24'-4" + 32'-0" + 24'-4" = 86'-8").

86.67' (86'-8") x .75 x 8 courses = 520 (12" x 8" x 16")
86.67' (86'-8") x .75 x 4 courses = 260 (8" x 8" x 16")
86.67' (86'-8") x .75 x 1 course = 65 (4" x 8" x 16" solid)

(b) The blocks displaced by the stepped footing are:

From Figure 5-5 2.25 (12" x 8" x 16" C.B.)
From Figure 5-6 36.75 (12" x 8" x 16" C.B.)
From Figure 5-7 139.50 (12" x 8" x 16" C.B.)
From Figure 5-7 3.00 (8" x 8" x 16" C.B.)
(Note: The 9th course is 8" x 8" x 16" blocks)

From Figure 5-8 103.00 (12" x 8" x 16" C.B.)
From Figure 5-8 6.00 (8" x 8" x 16" C.B.)
(The 9th course is 8" x 8" x 16")

The total blocks displaced will be:

281.50 (12" x 8" x 16")
9.00 (8" x 8" x 16")

(c) The masonry blocks required will be the total blocks computed in (a), less the displaced blocks computed in (b) above.

From (a)	520.0	(12" x 8" x 16")
From (b)	(281.5)	(12" x 8" x 16")
Net total	238.5	(12" x 8" x 16")
From (a)	260.0	(8" x 8" x 16")
From (b)	(9.0)	
Net total	251.0	(8" x 8" x 16")
From (a)	65.0	(4" x 8" x 16" solid)
From (b)	-0.-	
	65.0	(4" x 8" x 16" solid)

Note: The stepped footing (Figures 5-3, 5-5, 5-6, 5-7 and 5-8), rises above the baseline footing elevation where the foundation height is calculated and the blocks displaced are deducted from the totals. If the stepped footing drops below the baseline footing (*example:* a basement area on a sloping lot that requires a fill for part of the basement), the blocks for the stepped footings will be added to the total quantity of masonry units.

There will always be some waste in laying masonry blocks. Some of the blocks may be broken during unloading from the trucks; some will be broken by the helpers, and some will be broken by the masons while working with them. There is no set guideline to follow in allowing for this waste. Here again, cost records from previous jobs are very helpful in making this allowance for waste. To help offset this waste many builders and estimators do not deduct small openings for the windows and doors, and for the blocks overlapping at the corners. This waste may be from 4% to 8%, or even higher, depending on the conditions and the workmen. *In estimating masonry blocks always allow for some waste.*

The blocks that will be needed for the house foundation from the blueprints and Figure 4-1 are:

1286. . . . . .12" x 8" x 16"
665. . . . . . 8" x 8" x 16"
188. . . . . . 4" x 8" x 16" solid

(Enter these totals on Line 5.2 Cost Estimate Work Sheet)

The quantity of masonry blocks in the totals above include the following:

1. Window and door openings were computed as solid, and no allowance was made for the blocks overlapping at the corners. These displaced blocks, plus an allowance of 4% have been added for waste.

2. A total of 74 linear feet of 8" x 8" x 16" blocks has been added to use as fillers around the floor trusses.

3. The foundation wall between the basement area and the utility area will be 12 courses of 12" x 8" x 16" blocks plus 1 course of 4" x 8" x 16" solid blocks.

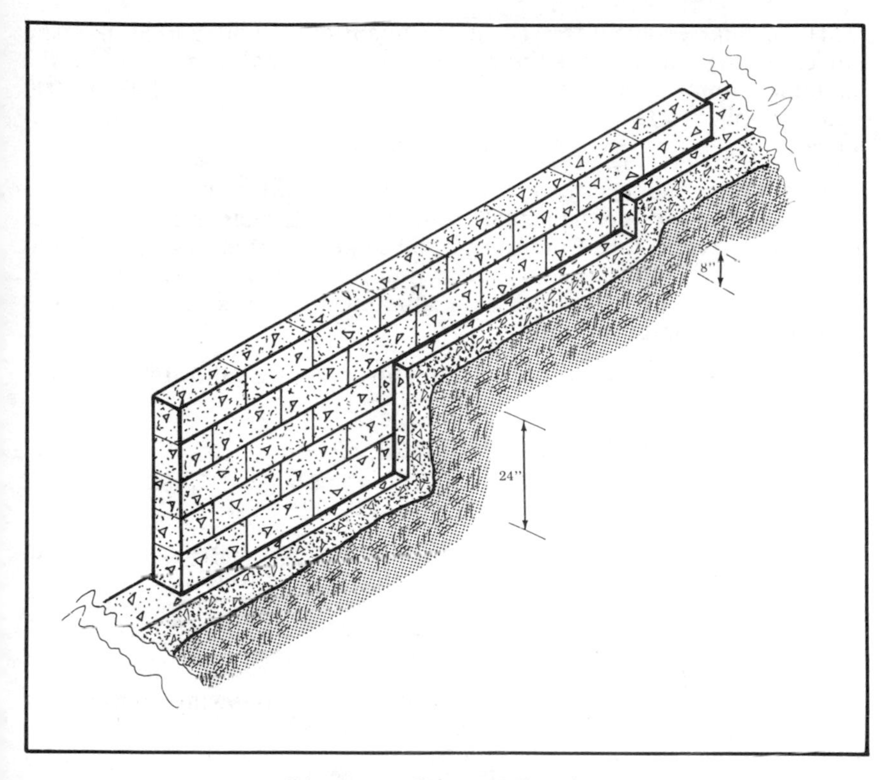

Blocks on Stepped Footing
Figure 5-9

4. The blocks displaced in the stepped footing around the garage and utility area have been deducted.

5. The supporting wall under the partition between the utility and garage area will be:

    1 course 8" x 8" x 16" C.B.
    1 course 4" x 8" x 16" solid C.B.

6. The retaining wall for the outside steps leading to the basement will be constructed with 12" x 8" x 16" blocks.

7. Corner and sash blocks are normally included in large orders by the supplier and it is not necessary to specify them. About 10% of the order is allowed for these blocks.

Estimating Mortar and Sand

There is a high percent of waste in mortar. It can be as much as 25% or more. Cost records from previous jobs will give the builder or estimator an idea of how much to allow for this waste. In addition to the waste the amount of mortar used by the masons depends on the mix and the thickness of the mortar joints. The 3/8" joint is the most common thickness, but it is not always used.

There will be some difference in the quantity of mortar used in laying different size blocks, but the difference is minimal and normally the amount of mortar and sand that will be estimated to lay 100 or 1000 blocks applies to 12" x 8" x 16", 8" x 8" x 16" and 4" x 8" x 16" combined.

Most tables showing the amount of mortar needed to lay 100 square feet of wall space, or the actual count of blocks such as 100 blocks, is calculated in cubic feet. For example, 60 cubic feet of mortar is needed to lay 1000 masonry blocks, but mortar is normally ordered by the number of bags of masonry cement and the amount of sand needed to produce this 60 cubic feet of mortar.

A formula that has proven reasonably accurate (allowing 25% for waste) is:

For 1 cubic foot mortar allow:
.50 bags masonry cement
100 lbs. sand

The amount of masonry cement and sand required to produce 60 cubic feet of mortar (allowing for waste) is calculated as follows:

60 x .50 (factor) = 30 bags masonry cement
60 x 100 (lbs.) = 6000 lbs. (3 Tons) sand

To lay 1000 masonry blocks estimate the following amount of masonry cement and sand (allowing 25% waste.).

30 bags masonry cement
3 tons sand

The number of blocks that will be required for the house foundation in 5.2 is:

	1286	12" x 8" x 16"
	665	8" x 8" x 16"
	188	4" x 8" x 16" solid
Total	2139	

The amount of mortar and sand required to lay these 2139 blocks is computed as follows:

$$\frac{2139}{1000} = 2.139$$

2.139 x 30 (bags cement per 1000 blocks) = 64.170 (or 64 bags)
*2.139 x 3 (tons sand per 1000 blocks) = 6.417 (or 6½ tons)
(Enter these totals on Line 5.3, Cost Estimate Work Sheet)

(*)It is cheaper to order sand by the truck load than in small quantities. If there will be brick or other masonry work later, the estimator will probably calculate the sand cost by the truck load. It will be less expensive this way.

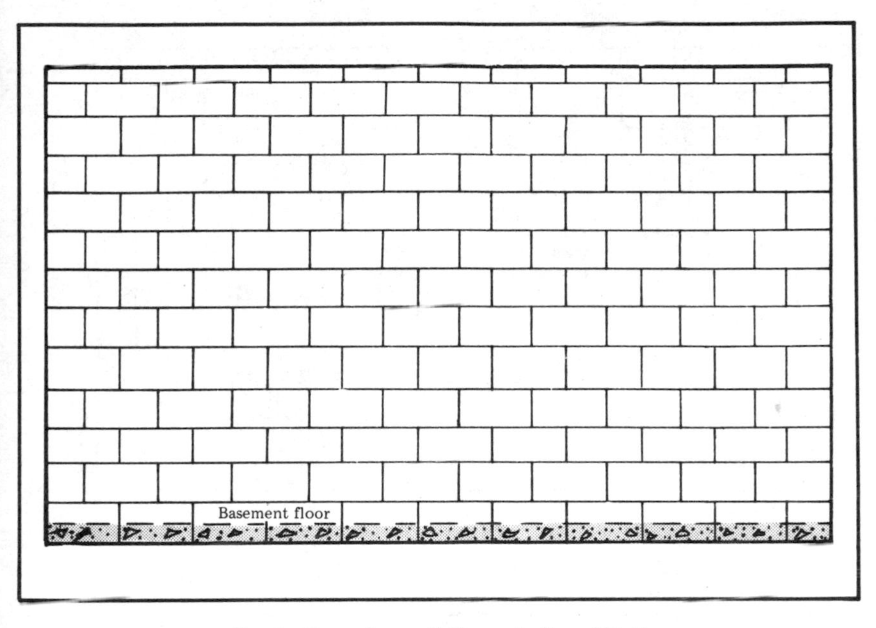

Scale Drawing of Foundation Wall
Figure 5-10

Basement Windows, Foundation Vents and Basement Doors

Basement Windows

Natural light and ventilation must be provided in all basements. Most basement windows are metal and are constructed with a flange so they will fit into the sash blocks that the mason installs around the window opening. This enables the mason to install the window before placing the lintel above the window opening. When it is planned to finish the interior walls of the basement, regular house-type windows will probably be installed. For these windows special provisions must be made for later installation. The method used by many builders is to allow an extra 3" in width and 1½" or 3" in height in the masonry opening for bucks onto which the finish jamb and head are attached. These bucks are secured to the masonry wall by bolts placed in the blocks by the masons during the construction of the foundation. (See the door section in Figure 5-11).

Figure 5-10, a scale drawing of a masonry wall, can be helpful in plotting the course on which the mason will start the masonry opening for the windows. The lintel for the height of most windows and doors in a basement rests on top of the 11th block course. When the masonry opening for the window is known, scale down from the lintel height the height of the window to get a quick visual picture of the window opening. You can tell at a glance on what course to start the window opening and how much space there will be for the window sill (Figure 5-11). If more space for the sill is needed, it may be necessary to leave out an extra course of blocks in the masonry opening.

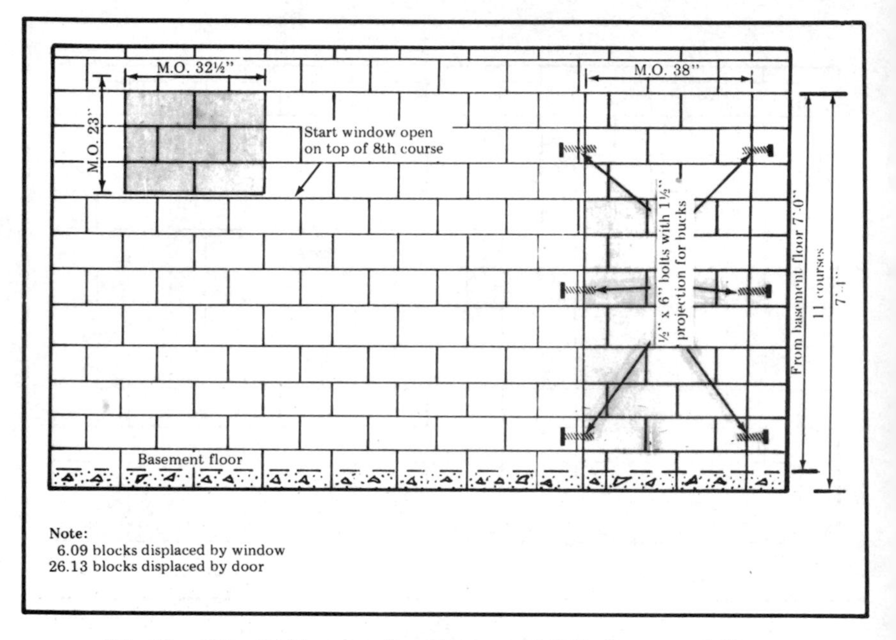

Plotting Block Courses for Basement Windows and Doors
Figure 5-11

Foundation Vents

In basementless spaces (crawl spaces), a minimum of four foundation wall vents should be provided. They should be located close to each corner.

Basement Doors

An outside entrance to and from the basement should be provided. The basement door should be a minimum of 2'-8" wide to accommodate most materials and supplies that will be moved to and from the basement area. Some builders set the door frame during the construction of the foundation, others allow an extra 3" in the width and 1½" in the height of the masonry opening for bucks (Figure 5-11). *Example:* 2/8 x 6/8 Basement Door:

Masonry Opening Width

Door	2'- 8"
Door frame (jamb thickness 1¼" x 2)	2½"
Bucks (1½" x 2)	3"
Clearance	½"
	2'-14" (3'-2") (or 38")

Masonry Opening Height

Door	6'- 8"
Door frame (jamb thickness 1¼" x 1)	1¼"
Buck (1½" x 1)	1½"
Clearance	½"
	6'-11¼"

The blueprints of the house from which the plot plan in Figure 2-1 was drawn specify the following windows and doors in the foundation wall: 2-15" x 20" (glass size) basement windows; 1-2/8 x 6/8 basement door; foundation vents, none.

(Enter on line 5.4 on the Cost Estimate Work Sheet)

Lintels, Beams, Column Posts, Anchor Bolts and Reinforcing Steel

Lintels

The lintels for masonry blocks come in precast reinforced and flat steel forms. The type of lintel used depends on the load it will carry and the span of the opening it will cover. For residential construction the masonry block lintel is usually adequate. There is a big difference in the price of the two lintels. A 48" masonry block lintel (flat steel) costs approximately $3.75 as compared to approximately $14.25 for a precast concrete lintel of the same size. This is nearly four times the price. If the precast lintel is needed, use it, otherwise use the block lintel. When there is a basement and crawl space in one foundation there will probably be an access door leading from the basement to the crawl space area. A lintel will be required for this access door. *Note:* Always order lintels a minimum of 8" longer than the masonry opening to allow a 4" bearing on each end.

Beams

Beams, or girders, are used to support loads. Beams may be laminated, box beams, built-up beams or steel. In residential construction the only beams that are normally required in the foundation are the beams that support the floor system, and occasionally the opening for the garage door that leads to the basement area.

The blueprints will show on the foundation plan what types of beams will be required. These beams are designed to support the load they must carry and the span between the column posts or piers. If the beam, or girder, is to be a built-up beam assembled by more than one framing member, it will be indicated on the foundation plan as a 3-2" x 10" wood girder, or a 3-2" x 12" wood girder (Figure 5-12). If the plans call for a steel beam it will be indicated by the size of the beam and the weight per foot (Figures 5-13 and 5-14). If the beam is supported by column posts, as in a basement, the plans will show the size of the posts (Figures 5-12 and 5-13). If the beam is supported by piers, as in a crawl space, the plans will specify the size of the piers (Figure 5-14).

Column Posts

The number and size of the column posts will be shown on the foundation plan as explained above and shown in Figures 5-12 and 5-13.

Anchor Bolts

These bolts are used to secure the dwelling to the foundation. The size to use and their placement in the foundation wall is not a matter of guess work. The bolts should be ½" x 16" with washers, and the spacing between bolts should not exceed 8'-0". In some areas the maximum spacing

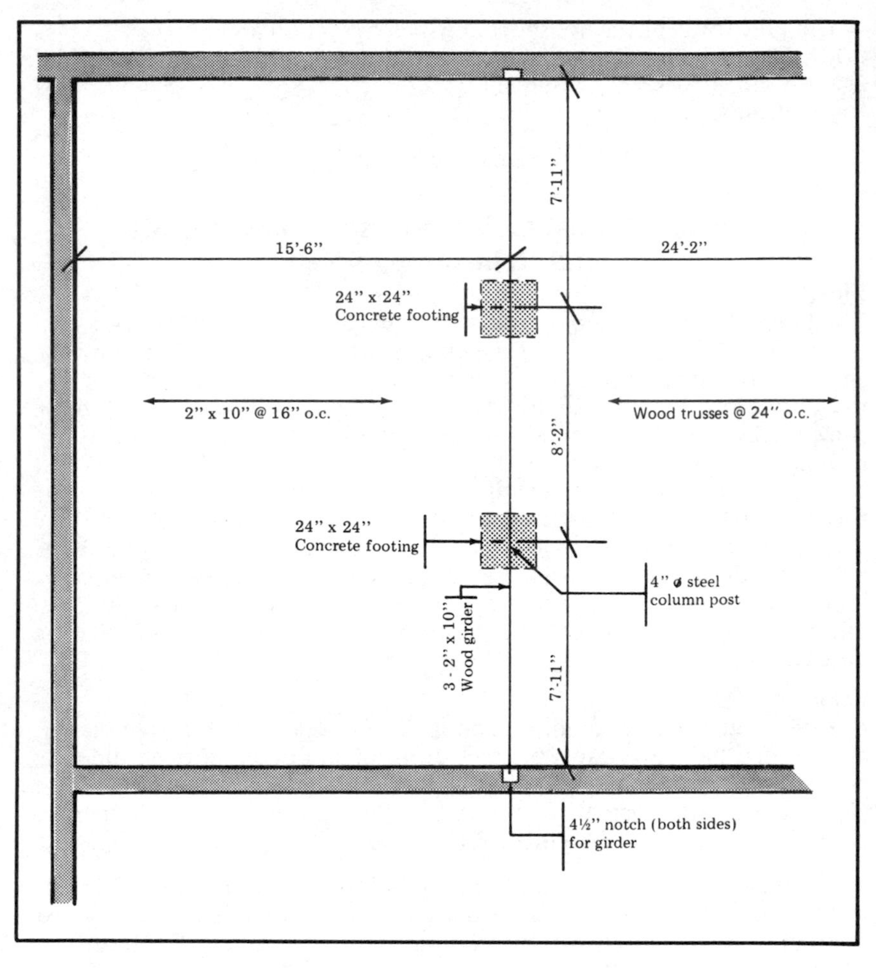

Plan Indicating Wood Girder
Figure 5-12

between bolts should not exceed 6'-0''. Not less than two bolts should be in each sill plate and the end bolts should be not more than 1'-0'' from the end of the sill (Figure 5-15).

To estimate the number of anchor bolts for a foundation wall:

1. Divide the perimeter of the foundation wall by 8 (if 8'-0'' spacing is to be used).

2. Add one bolt for each exterior and interior corner in the masonry wall.

3. Add 1 and 2 above for the total number of anchor bolts required. ***Example:***

(a) The perimeter of a house foundation wall is 192'-0''

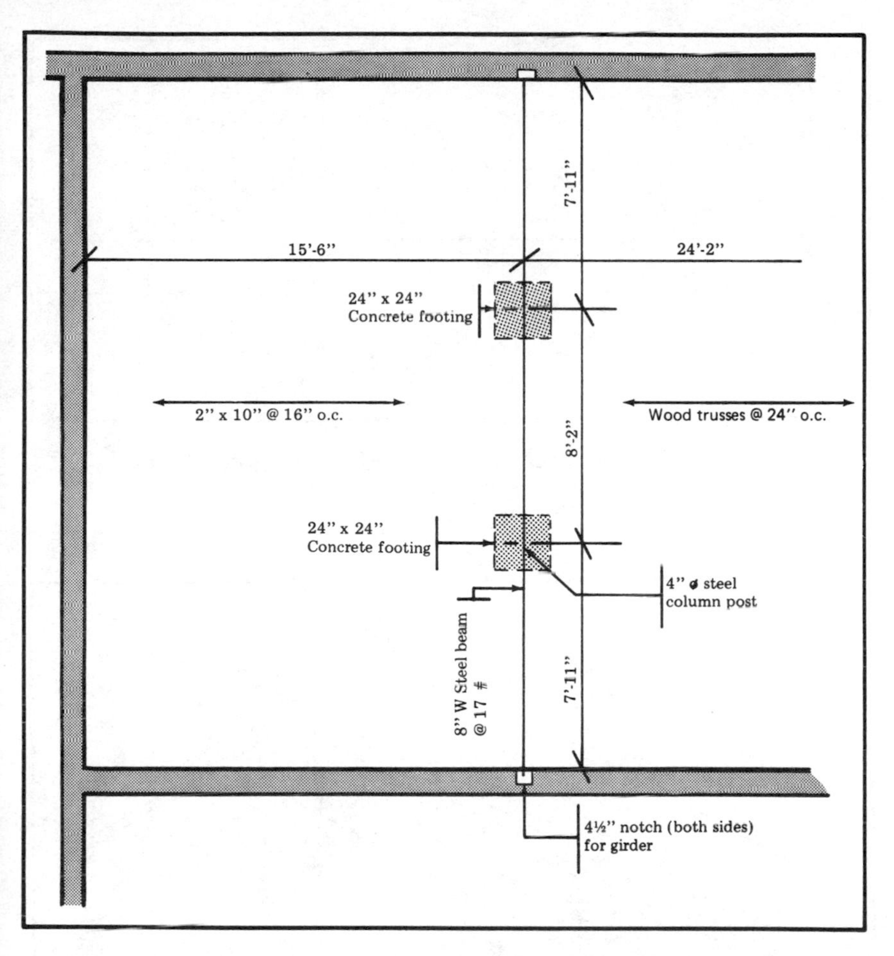

Plan Indicating Steel Beam with Steel Column Posts
Figure 5-13

$$\frac{192}{8} = 24 \text{ anchor bolts}$$

(b) There are six exterior and interior corners

$$6 \times 1 = 6 \text{ anchor bolts}$$

(c) There will be a total of 30 anchor bolts needed for this foundation (24 + 6 = 30).

Reinforcing Steel

In some areas reinforcement for the masonry foundation walls may be required. The local building codes or the blueprints, or both, will detail the size and spacing of the reinforcement required.

The material take-off for lintels, beams, column posts, anchor bolts and reinforcing steel for the cost estimate work sheet will be:

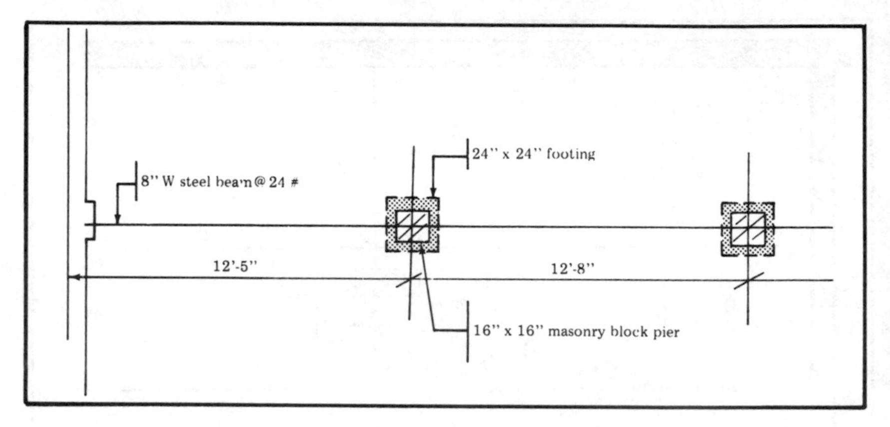

Plan Indicating Steel Beam with Masonry Block Piers
Figure 5-14

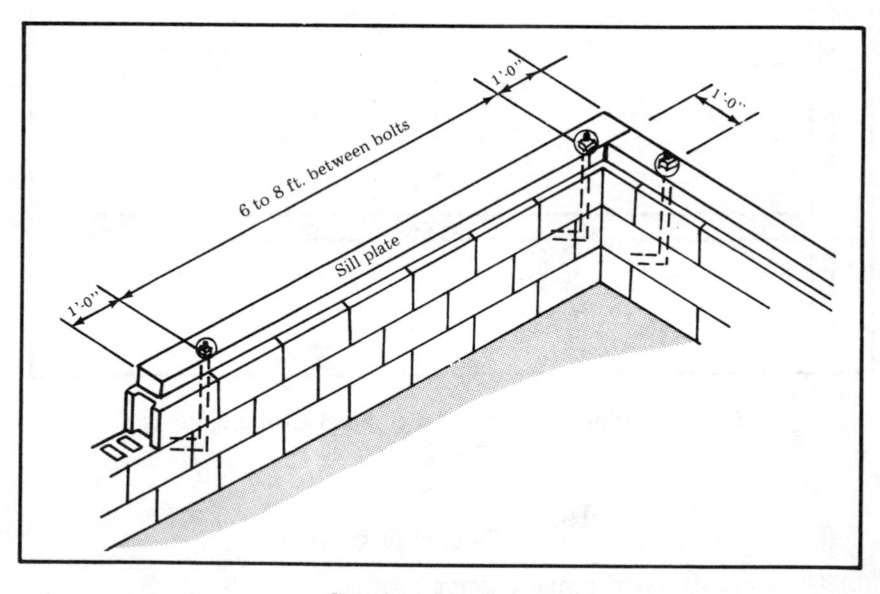

Spacing Anchor Bolts
Figure 5-15

Lintels: 3-48" block lintels (2 basement windows with masonry openings of 32½" each, and 1 door masonry opening of 38". Note: There are no access doors).

Beams: This is a factory-built house and the floor system (including the preassembled wood girder) is included in the package price. If the girder had not been preassembled, the labor that would have been required to fabricate it would have been estimated at two man-hours per 10 linear feet. This would include fabricating the beam, placing it and leveling and aligning it.

Column Posts: 2 - 4'' x 8'-0'' diameter steel posts.

Anchor Bolts: 30 - ½'' x 16'' w/washers.

Reinforcing Steel: None.

(Enter on Line 5.5 on the Cost Estimate Work Sheet)

Waterproofing and Drain Tile

Waterproofing

Basements and habitable spaces below grade must be protected against moisture. Waterproofing keeps the water standing around the foundation walls from penetrating.

The most common method of waterproofing used in residential construction is parging the foundation walls below grade with ½'' of masonry mortar and bituminous coating (Figure 5-16).

In locations where the foundation is subjected to a high water table or where surface drainage is a problem, a membrane of hot tar and built-up fabric should be used. This work is normally done by subcontractors and is seldom required in residential construction.

To estimate the material for ½'' of parging on a foundation wall:

1. Multiply the total square feet of the foundation area that is to be parged by .00154 to get the number of cubic yards of mortar mix required (allow 5% for waste).

2. Allow 3000 pounds of sand and 12 bags of masonry cement per 1 cubic yard of mortar.

Parging the foundation walls for the house in Figure 3-2, including the perimeter of the walls around the garage and utility area where stepped footing is required (see Figures 5-5, 5-6, 5-7 and 5-8) will require 18 bags of masonry cement and 2.19 tons of sand. It was computed as follows:

1. The parging will extend to the top of the 8th course of blocks (64'') and will overlap on the footers. Allowing for the footing overlap, a height of 6'-0'' will be used for the height of the parging.

2. The basement wall is 105'-4'' in length (39'-8'' + 26'-0'' + 39'-8'' = 105'-4''). The number of square feet for the parging on the basement foundation walls will be: 6.0' x 105.33' (105'-4'') = *631.98 square feet* (no deduction is made for the door).

3. For the perimeter of the garage and utility area the calculations are: *From Figure 5-5 (right side offset)* 6.0' x 6.0' = 36.0' (less displaced block area) 36.0 less 2.0' = *34.0 square feet.*

From Figure 5-6 (garage front) 6.0' x 24.33' (24'-4'') = 145.98 square feet (less displaced block area) 145.98 less 32.70 = *113.28 square feet.*

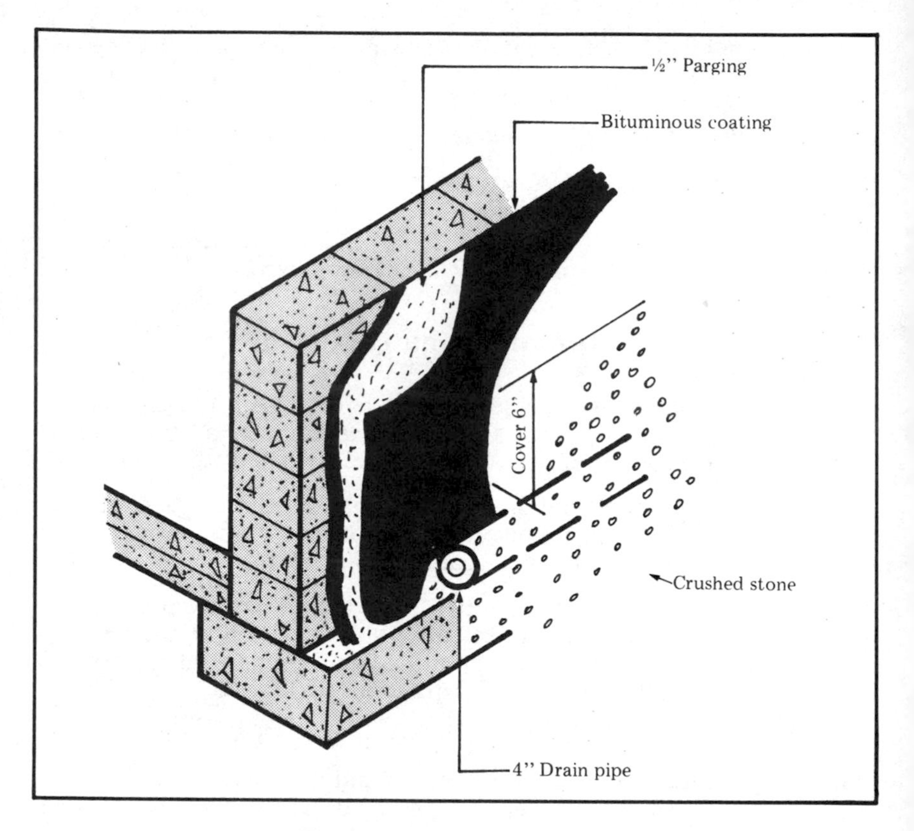

Foundation Waterproofing
Figure 5-16

From Figure 5-7 (left side) 6.0' x 32.0' = 192.0 sq. ft. (less displaced block area) 192.0 less 123.90 = *68.10 square feet.*

From Figure 5-8 (back) 6.0' x 24.33' (24'-4") = 145.98 (less displaced block area) 145.98 less 91.64 = *54.34 square feet.*

The total square feet of parging will be:

Basement walls	631.98 square feet
From Figure 5-5	34.00 square feet
From Figure 5-6	113.28 square feet
From Figure 5-7	68.10 square feet
From Figure 5-8	54.34 square feet
Total	901.70 square feet

The number of cubic yards of mortar will be: 901.70 square feet x .00154 (factor) = 1.39 cubic yards.

The amount of masonry cement and sand that will be required for 1.39 cubic yards of mortar will be:

Masonry cement
(Formula: 12 bags per 1 cubic yard)

1.39 x 12	=	16.68
(5% for waste)	=	83
		17.51 or *18 bags masonry cement*

Sand: (Formula: 3000 pounds per 1 cubic yard)

1.39 x 3000	=	4170.0
(5% for waste)	=	208.5
		4378.5 lbs. or *2.19 tons*

*The labor for parging the foundation walls can be estimated at:

2 man-hours per 100 square feet

Example: $\frac{901.70 \text{ square feet}}{100} = 9.02$

9.02 x 2 = 18.04 or *18 man-hours*

The amount of bituminous coating that will be required is computed as follows:

Formula: 1 gallon bituminous coating per 100 square feet.

$\frac{901.70 \text{ square feet}}{100} = 9.02$

9.02 x 1 = 9.02 gallons

Note: Bituminous coating is normally sold in 5 gallon cans. Order *2-5 gallon cans.*

The labor to brush on the bituminous coating can be estimated at:

1½ man-hours per 100 sq. ft.

$\frac{901.70 \text{ square feet}}{100} = 9.02$

9.02 x 1.5 (1½) = 13.53 or *14 man-hours*

Drain Tile

Footing drain tile should be installed around foundations enclosing basements or habitable spaces below grade. The drain tile relieves the hydrostatic pressure against the foundation wall and basement floor. The tile must discharge by gravity or by pump to an outfall, such as a drainage ditch or into a sump pit from which it can be pumped into a storm or sanitary drainage system. Drainage pipe can be either 4'' farm tile (12'' long), 4'' perforated bituminous fiber or 4'' perforated plastic (10'-0'' in length). There should be a minimum of 2'' of crushed stone under the pipe and a minimum of 6'' of crushed stone to cover it (Figure 5-16).

From the plot plan in Figure 2-1 the grade elevation at the lowest corner of the lot is 98.26' (left front corner) and the basement footing elevation is 95.90' (Figures 5-5 and 5-8). The basement footing where the drain tile will be is 2.36' (2'-4'') lower than the lowest grade elevation on the lot so the discharge from the drain tile will have to be pumped into the sanitary drainage system. This will be no problem because there will be a pump and sump pit for the basement floor drain and the discharge from the drain tile can discharge into this sump pit.

The material required for the drain pipe will be:

105'-4" (39'-8" + 26'-0" + 39'-8" = 105'-4")

$$\frac{105.33'}{10 \text{ (length of pipe)}} = 10.53 \text{ or } 11 \text{ joints}$$

11 joints @ 10' = *110'-4" perforated plastic pipe*

The factor for computing the cubic yards of crushed stone 18'' wide and 12'' deep (after deducting for the volume of the 4'' drain pipe) is:

cubic yards = .05232 x lineal feet

.05232 x 105.33' = 5.51 cubic yards

Proof

18" wide x 12" deep x 105'-4" length

1.5' (18") x 1.0' (12") x 105.33'	=	158.00 cubic feet
(Less volume of drain tile)	=	(9.19) cubic feet
(Formula: $\pi \times r^2 \times$ length)		148.81 cubic feet

$$\frac{148.81}{27} = 5.51 \text{ cubic yards}$$

1 cubic yard crushed stone = 2700 pounds

5.51 cu. yards x 2700 = 14,877 lbs. or 7.44 tons (rounded off) *7½ tons*

The amount of material for the drain tile around the basement will be: 110'-4'' perforated plastic pipe, 2-90° ells, and 7½-tons crushed stone.

(Enter on Line 5.6 Cost Estimate Work Sheet)

The labor to install this drain tile, including the gravel can be estimated at: 10 man-hours per 100 linear feet.

$$\text{Example: } \frac{105.33'}{100} = 1.05$$

1.05 x 10 = 10.50 or *11 man-hours*

Miscellaneous Materials

Some of the miscellaneous materials for the foundation that may be overlooked by the estimator are:

1. *Wall ties:* If brick is to be used and they will start below grade, wall ties will be needed in the foundation walls.

2. *Polyethylene:* This film is used for vapor barriers, covering masonry cement and other materials.

3. *Bucks for basement windows and doors:* 2'' x 4'' or 2'' x 6'' framing lumber is the most common used material for bucks.

4. *Bolts:* ½'' x 6'' bolts with washers are used to secure the bucks to the masonry walls (Figure 5-11).

5. *Door frame for basement door:* If the door frame is to be set before the masonry blocks are laid the door frame may be included here, otherwise include it later with the exterior finish material.

6. *Areaways:* Windows and vents below grade will have to be protected by areaways.

7. *Dryer vents:* If the laundry area will be in the basement, install the dryer vent during the construction of the foundation.

8. *Mortar antifreeze:* During cold weather it may be required.

9. *Termite protection:* This can be by metal shields or chemical treatment.

10. *Brushes for the bituminous coating:* if needed.

11. *Scaffold rental:* if needed.

The miscellaneous materials to be included in this cost estimate for the foundation are:

Item	Cost
1 - box wall ties (500)	$12.42
1 - roll polyethylene (12' x 100')	15.20
1 - 2'' x 6'' x 12' (door buck)	4.08
1 - 2'' x 6'' x 8' (door buck)	2.72
6 - ½'' x 6'' bolts w/washers @ .35¢	2.10
Total	$36.52

(Enter on Line 5.7 Cost Estimate Work Sheet)

Note: Two areaways will be required for the basement windows. These will be constructed from brick so the cost will be included with the brick estimate.

Masonry Labor

Masons may charge an hourly rate or they may subcontract the foundation blocks by the job, or by the block.

If an hourly wage is paid to the masons and their helpers, the builder will have the additional expense of paying the F.I.C.A., F.U.T.A., Worker's Compensation and Liability insurance and taxes on their wages. The amount paid on these taxes and insurance premiums is a percentage of each employee's earnings. There will also be the administrative expense of keeping the payroll records and filing the tax returns — expenses that must not be overlooked by the builder or estimator.

If the mason is a masonry contractor, he will hire his own masons and helpers and the responsibility of the payroll records, taxes and insurance is his. He will contract the masonry blocks either by the job or by the block. *Note:* If the masonry work is let as a subcontract, get a copy of the masonry contractors worker's compensation policy before any work is started.

Many factors control the number of masonry blocks a mason can lay in one day. The size of the blocks, the number of openings in the foundation walls, the height of the walls and the efficiency of the workmen are some of these factors. Some masons will average 200 or more blocks in one 8 hour shift, others may average half this number. If there is an option of paying the masons by the hour or by contract, weigh all of the factors mentioned above before making a final decision.

In the cost estimate for the masonry blocks in the foundation for the house used in this book (Figure 4-1), the masonry work was let on a contract for .75¢ per block including all sizes. The number of blocks estimated for this foundation is:

	1286	12" x 8" x 16"
	665	8" x 8" x 16"
	188	4" x 8" x 16" solid
Total	2139	masonry blocks

The masonry labor cost will be:

2139 blocks @ .75¢ = $1604.25.

(Enter on Line 5.8 Cost Estimate Work Sheet)

Other Labor

In addition to the masonry labor for the foundation there will be some other labor that is normally performed by the regular employees of the builder. The payroll taxes and insurance for these workmen will be computed on the total estimated payroll for the house that is shown later in this book.

The other labor costs for this foundation are:

1. Labor to spread the crushed stone in the basement.
   12 man-hours @ $6.25 .... $75.00
2. Labor to parge foundation walls.
   18 man-hours @ $6.25 .... 112.50
3. Labor to brush on bituminous coating.
   14 man-hours @ $6.25 .... 87.50
4. Labor to install drain tile.
   11 man-hours @ $6.25 .... 68.75

Total $343.75

(Enter on Line 5.9 Cost Estimate Work Sheet)

Cost Estimate Work Sheet For Foundation

5.1 Stone fill under concrete slab:
_____ Tons stone @ _____ = $_________

5.2 Masonry blocks:
_____ 12" x 8" x 16" @ _____ = = _________
_____ 8" x 8" x 16" @ _____ = _________
_____ Total blocks $ _________

5.3 Mortar and sand:
_____ Bags masonry cement @ _____ $ _________
_____ Tons sand @ _____ = _________
$ _________ _______

5.4 Basement windows, foundation vents and basement doors:
_____ Basement windows (size _____) @ _____ = $_________
_____ Foundation vents (size _____) @ _____ = _________
_____ Basement doors (size _____) @ _____ = _________
$ _________ _______

5.5 Lintels, beams, column posts, anchor bolts, and reinforcing steel:
_____ Lintels (size ___) @ _____ = $ _________
_____ Beams (size ___) @ _____ = _________
(If necessary itemize on separate sheet and enter total cost here)
_____ Column posts (size _____) @ _____ = _________
_____ Anchor bolts (size _____) @ _____ = _________
_____ Reinforcing steel (size _____) @ _____ = _________
$ _________ _______

5.6 Waterproofing and drain tile:
_____ Bags masonry cement @ _____ $ _________

Total for page 1 (carry forward) $ _______

Brought forward from page 1 $ ________

____ Tons sand @ ____ = ________

____ (5-gal. cans) Bituminous coating @ ____ = ________

____ Lin. ft. drain tile @ ____ = ________

____ Ells @ ____ = ________

____ Tons stone @ ____ = ________

$ ________ ________

5.7 Miscellaneous materials:
(Itemize on separate sheet and enter totals here

Cost of material $ ________

Sales tax (____%) ________

Total cost of material $________(1)

5.8 Masonry labor:
________ Blocks @ ________ = $ ________(2)

5.9 Other labor:
(Itemize on separate sheet and enter totals here) $ ________(3)

(Add lines (1), (2) and (3) Cost of foundation $ ________]
(Enter on Line 5 Form 100)

Chapter 6

Floor Systems

The estimator will receive most of the information he needs on the floor system from the wall section of the blueprints. This section will show the size of the sill plate, the size and spacing of the floor joists and the type and thickness of the subfloor. Some plans also show the size of the girders.

Framing lumber is normally priced and sold by the thousand board feet and is abbreviated as follows:

B.F. or bd. ft.	Board foot
B.M. or b.m.	Board (foot) measure
M.b.m. or Mfbm	Thousand (feet) board measure

Board measure is the standard basis of measuring lumber and the board foot is the unit of measurement. A board foot of lumber is 1" thick x 12" wide x 1'0" long (one square foot, 1" thick). (Figure 6-1). The calculations for computing board measure are:

(Nominal)

$$\frac{\text{Thickness (inches) x width (inches) x length (feet)}}{12}$$

Example: How many board feet are in a piece of lumber 1" x 12" x 1'-0"?

Answer: $\frac{1\text{" x } 12\text{" x } 1\text{'-0"}}{12}$ = 1 board foot

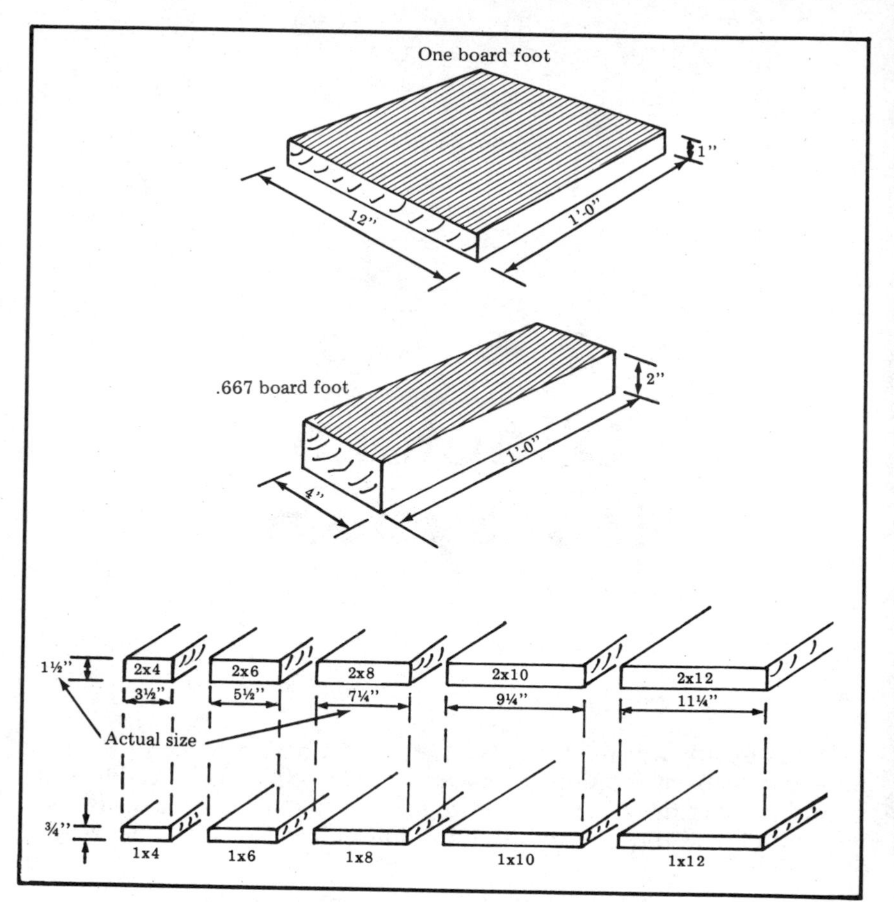

Nominal and Actual Lumber Sizes
Figure 6-1

How many board feet are in a piece of lumber 2" x 4" x 1'-0"?

Answer: $\frac{2" \times 4" \times 1'\text{-}0"}{12}$ = .667 board feet

1 — 2" x 8" x 14' = 18.667 board feet

Solution: $\frac{2" \times 8" \times 14'}{12}$ = 18.667 board feet

1 — 2" x 10" x 16' = 26.667 board feet

Solution: $\frac{2" \times 10" \times 16'}{12}$ = 26.667 board feet

1 — 2" x 12" x 18' = 36.000 board feet

Solution: $\frac{2" \times 12" \times 18'}{12}$ = 36.000 board feet

Lumber dealers often price sheathing by the hundred lineal feet (c/lin. ft.). To convert lineal feet to board measure do the following:

Example: How many board feet are in 1250 lin. ft. of 1" x 8"?

Answer: $\dfrac{1'' \times 8'' \times 1250'}{12} = 833.333$ board feet

To convert board measure to linear feet multiply board feet by 12 and divide by the thickness and width.

Example: How many linear feet are in 833.333 board feet of 1" x 8"?

Answer: $\dfrac{833.333 \times 12}{1'' \times 8''} = 1250$ lin. ft.

The normal method of pricing lumber is by the thousand board feet (M.b.m.). However, many lumber dealers price and sell framing by the piece. *Example:* 2" x 4" x 8' @ $1.66 each, 2" x 8" x 10' @ $4.60 each, etc. If you need to convert these unit prices into the price per M.b.m. do the following:

1. Divide the number of board feet in each piece of framing into the unit cost.

2. Multiply the answer in 1 by one thousand. *Example:* If 2" x 4" x 8's are priced at $1.66, what is the price per M.b.m.?

(1 — 2" x 4" x 8' = 5.333 board feet)

1. $\dfrac{\$1.66}{5.333} = .311$ cents per board foot

2. .311 cents x 1000 = $311.00 M.b.m.

Example: If 2" x 8" x 10's are priced at $4.60 each what is the price per M.b.m.?

(1 — 2" x 8" x 10' = 13.333 board feet

1. $\dfrac{\$4.60}{13.333} = .345$ cents per board foot

2. .345 cents x 1000 = $345.000 M.b.m.

There are many lumber tables showing the board feet in various sizes of lumber. Every estimator should keep one of these tables at his desk. However, with the electronic calculators available it is much faster to estimate the number of board feet, and the total cost of a lumber order by direct calculations.

Example: What will the lumber cost for the following?

200 — 2" x 4" x 8' @ $311.00 per M.b.m.
65 — 2" x 6" x 14' @ 340.00 per M.b.m.
78 — 2" x 10" x 16' @ 355.00 per M.b.m.

Answer: $\frac{200 \text{ x } 2\text{"} \text{ x } 4\text{"} \text{ x } 8\text{'}}{(12 \text{ x } 1000)}$ @ \$311.00 M.b.m

$\frac{12800}{12000}$ = 1.067 x \$311.000 = \$331.84

$\frac{65 \text{ x } 2\text{"} \text{ x } 6\text{"} \text{ x } 14\text{'}}{(12 \text{ x } 1000)}$ @ \$340.00 M.b.m.

$\frac{10920}{12000}$ = .910 x \$340.00 = 309.40

$\frac{78 \text{ x } 2\text{"} \text{ x } 10\text{"} \text{ x } 16\text{'}}{(12 \text{ x } 1000)}$ @ \$355.00 M.b.m.

$\frac{24960}{12000}$ = 2.08 x \$355.00 = 738.40

Total \$1379.64

Sill Plate, Girder, Ledger or Joist Hangers

Sill Plate

The sill plate, or mud sill, attaches the floor system to the foundation with anchor bolts (Figure 5-15). The minimum thickness of the plate is 2 inches (nominal measurement) and the minimum width as required must be not less than 1½'' bearing for the ends of the joists. No joints should occur over window and door openings.

To estimate the material for the sill plate:

1. Check the wall section of the blueprints for the size of the sill plate. In residential construction it is usually 2'' x 6''.

2. Add the total linear feet of the foundation wall where the sill plate will be installed.

3. Divide the total linear feet by the length desired for each piece of framing lumber for the number of pieces required. *Example:* If 12'-0'' lengths are desired, divide by 12; if 14'-0'' lengths are desired, divide by 14.

4. If the quantity of board feet in the sill plate is needed, convert the number of pieces to board feet as explained at the beginning of this chapter.

Example: How many pieces of framing lumber, and how many board feet will be required for the foundation sill plate in Figure 6-2?

1. The blueprints show the sill plate to be 2'' x 6''.

2. The total lineal feet of the foundation walls where the sill plate is required is 242'-0''.

3. The plate will be 2'' x 6'' x 12'. The number of pieces required (allowing for waste) will be 21 - 2'' x 6'' x 12' (242.0' divided by 12 = 20.17 or 21 pieces.

4. The quantity of board feet will be:

$$\frac{21 \text{ x } 2\text{"} \text{ x } 6\text{"} \text{ x } 12\text{'}}{12} = 252 \text{ bd. ft.}$$

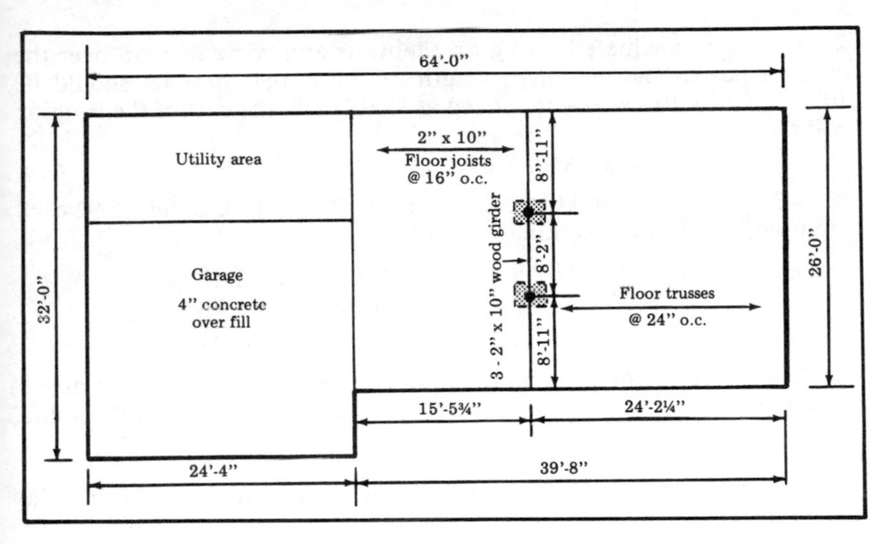

Foundation Plan
Figure 6-2

Girder
The bearing on walls should be a minimum of 4 inches, with a minimum of 6 inches of solid masonry supporting the girder. A ¼" bearing plate may be required where the girder rests on masonry blocks. All joints of a wood girder must be made over the column supports. Built-up girders of three or more members should be nailed from both sides as follows:

1. Two 20d nails at ends of piece and each splice.
2. Two rows of 20d nails in between splices at top and bottom of girder 32 inches o.c., with staggered nails.

The span of the girder should be in accordance with standard engineering practice based upon the allowable fiber stress of the lumber. The size of the girder shown on the blueprints will be designed in accordance with the maximum allowable span for the house shown. The type and size of the girder should never be down-graded from the size shown (Figure 6-2).

To estimate the material for girders:

1. Check blueprints for size of girder and spans between column posts or piers.

2. Add the total length of the girder and calculate the most practical length of material (keep in mind all joints must be made over the column supports).

Example:
1. From Figure 6-2 the size of the girder is 3 - 2" x 10", and the column posts are 8'11" from each outside wall to the first column post and 8'2" between the two column posts.

2. Allow for adequate bearing on the walls and some overlap over the column posts. The minimum length of the girder material should be 10'0''. If 2'' x 10'' x 10's are priced at $5.92 each, the cost of the framing will be:

9 - 2'' x 10'' x 10' @ $5.92 = $53.28

If 3 - 2'' x 10'' x 18' @ $10.75 each, and 3 - 2'' x 10'' x 10' @ $5.92 each are used, the cost of the lumber will be:

3 — 2'' x 10'' x 18' @ $10.75	=	$32.25
3 — 2'' x 10'' x 10' @ 5.92	=	17.76
Total		$50.01

There will be some waste in the lumber regardless of the length of the girder material used, but using fewer pieces requires fewer splices and results in less labor costs.

There are approximately 30 - 20d nails per pound. Following the recommended nailing schedule, it will require 1½ lbs. 20d nails for the girder.

Ledger

When floor joists frame into the side of a wood girder as shown in Figure 6-3, use either steel joist hangers or wood ledger strips. If wood ledger strips are used they must be at least 2'' x 2'', and must be nailed to the girder with three 16d nails at each joist.

If 2'' x 2'' wood ledger strips are used, 2'' x 4'' can be ripped in half to make them. *Example:* If the length of the girder is 24'-0'' the material for the ledger strips will be:

2 - 2'' x 4'' x 12' (ripped in half = 48' ledger strips)

There are approximately 54 - 16d nails per pound. Allow 2 - pounds of 16d nails for 24'-0'' of ledger strips.

The material for the sill plate, girder and ledger will be:

Sill plate	21 — 2'' x 6'' x 12'	(252 bd. ft.)
(*) Girder	3 — 2'' x 10'' x 18'	(90 bd. ft.)
	3 — 2'' x 10'' x 10'	(50 bd. ft.)
Ledger	2 — 2'' x 4'' x 12'	(16 bd. ft.)
Nails	2 lbs. 16d common	
	1½ lbs. 20d common	

(*) Enter on Line 6.1 Cost Estimate Work Sheet)

(*) (1) If the girder was included as the beam in the foundation cost work sheet do not enter it here.
(2) If the house is factory built, part or all of the floor system may be included in the package price of the house. Omit any material that is included in the house package price. It will be included in the house package cost in the next chapter.

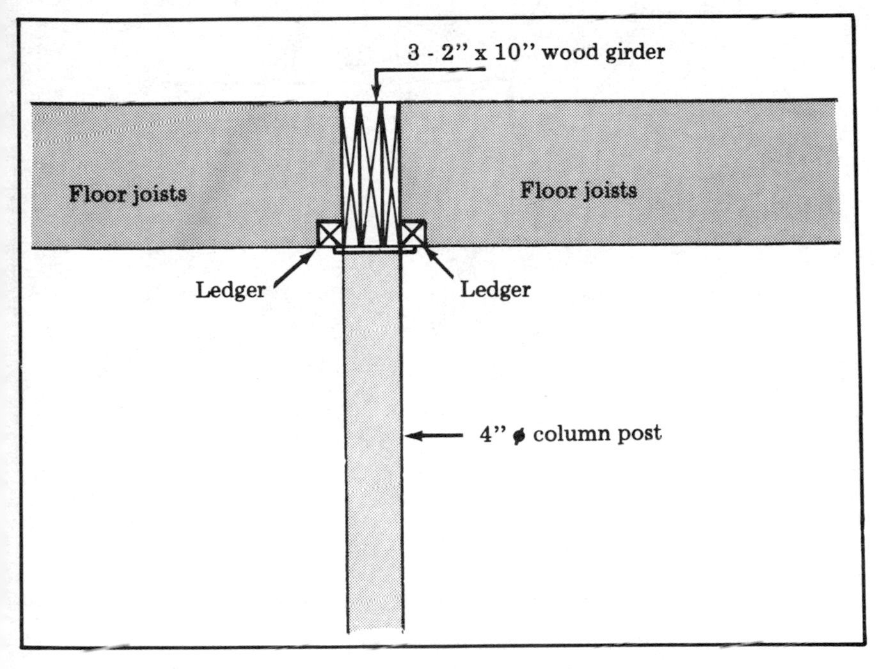

Floor Joists Framing into Girder
Figure 6-3

Floor Joists
Span of joists is considered to be clear span between inner faces of supports. They should not exceed the allowable design stresses using standard engineering analysis procedures. It is beyond the scope of this book to list the tables of maximum allowable spans for all wood floor joists. The designer of the house will show on the blueprints the size and spacing of the floor joists for each span, (Figure 6-2). *The size and spacing of the joists shown on the plans should never be downgraded.*

The joists must have a minimum bearing on the exterior wood sill plate of 1½ inches. A continuous 2 inch header joist (band joist) should be installed to prevent lateral movement and to provide subfloor nailing, (Figure 6-7). Toe nail floor joists to the sill plate with three 8d nails. The header should be toe nailed to the sill plate with 8d nails spaced 16 inches o.c. and the header should be nailed to each floor joist with a minimum of two 16d nails.

Floor joists framing into the side of a wood girder or beam should be supported by a ledger strip (Figure 6-3), and toe nailed to the girder with three 10d or 16d nails.

Floor joists framing over girders and bearing partitions must have a minimum lap of 4 inches (Figure 6-4) and be nailed together with three 16d nails. The joists should be nailed to the girder or bearing partition with three 8d toe nails.

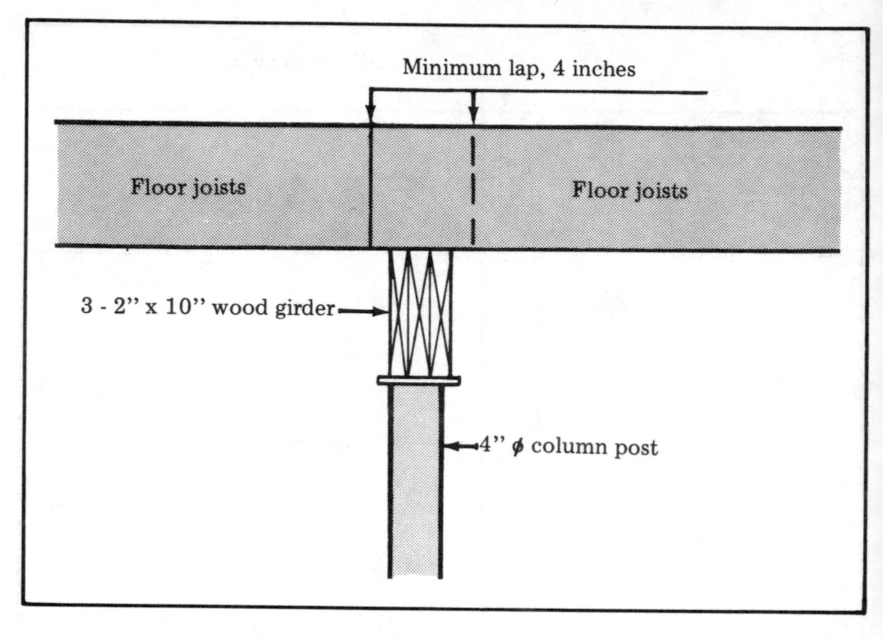

Floor Joists Resting on Top of Girder
Figure 6-4

Floor joists should be doubled under all parallel partitions and nailed together top and bottom every 32 inches with 10d or 16d nails.

When estimating the number of floor joists from the blueprints, take-off the number of pieces needed for each section separately, (Figures 6-5 and 6-6). If the joists are spaced at 16" o.c. multiply the lineal feet perpendicular to the length of the joists in each section by .75 (75%). To this number add one joist for the starter and one for each double joist. Do not deduct joists for small openings-the cut-outs can be used for headers.

Example: What length and how many floor joists will be required for the section of the first floor with the stair opening in Figure 6-5? Joist spacing is 16" o.c.

Solution:

1. The total lineal feet perpendicular to the length of the joists is 26'-0".
2. The span between supports is 15'-5¾" requiring 16' lengths.
3. The number of 2" x 10" x 16' floor joists that will be required for this section will be:

26.0' x .75 =	19.50
Starter	1.00
Double joists (2)	2.00
	22.50 or 23 joists

Total: 23 — 2" x 10" x 16'

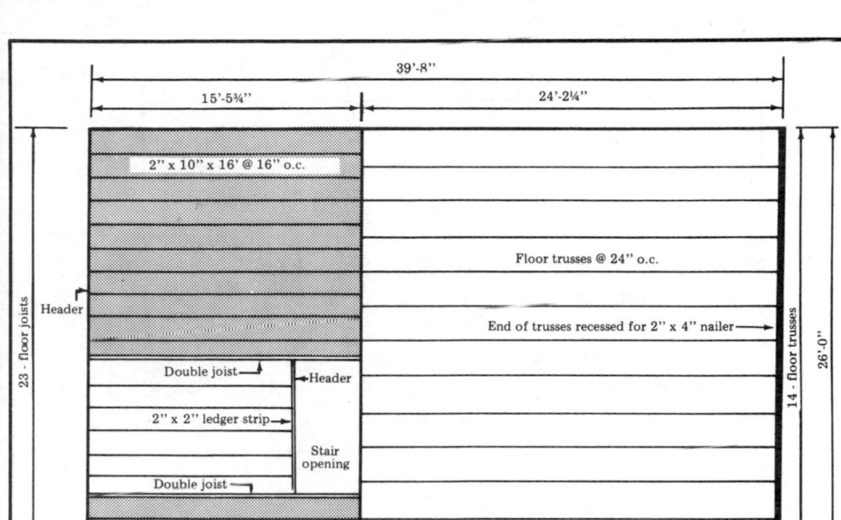

Figure 6-5
First Floor

If floor joists or floor trusses are spaced 24" o.c., multiply the lineal feet perpendicular to the joists by .50 (50%) and add one joist for the starter and one for each double joist.

Example: In Figure 6-5 there are two floor elevations (not shown) on the first floor. Floor trusses were designed by the architect to provide the two elevations, and also for the 24'-2¼" span. The floor trusses are on 24" o.c. and the number required is:

1. The total lineal feet perpendicular to the trusses is 26'-0"

2. 26.0' x .50 = 13
   Starter 1
   ———
   14

   Total: 14 — Floor trusses

The calculations for the floor joists on the second floor (Figure 6-6) are:

1. *Back section:*
   40.0' x .75 = 30
   Starter 1
   Double joists (5) 5
   ———
   36

   Total: 36 — 2" x 10" x 12'

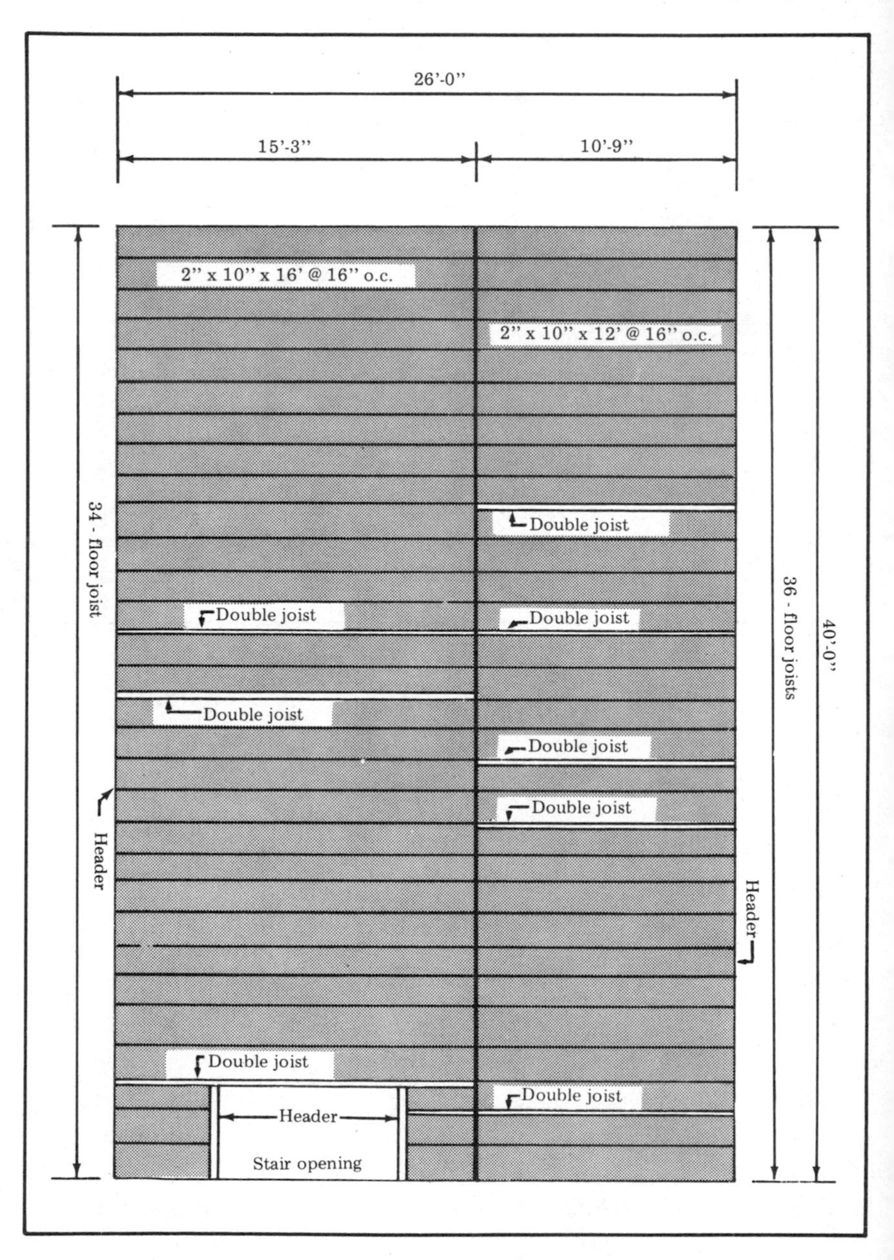

Figure 6-6
Second Floor

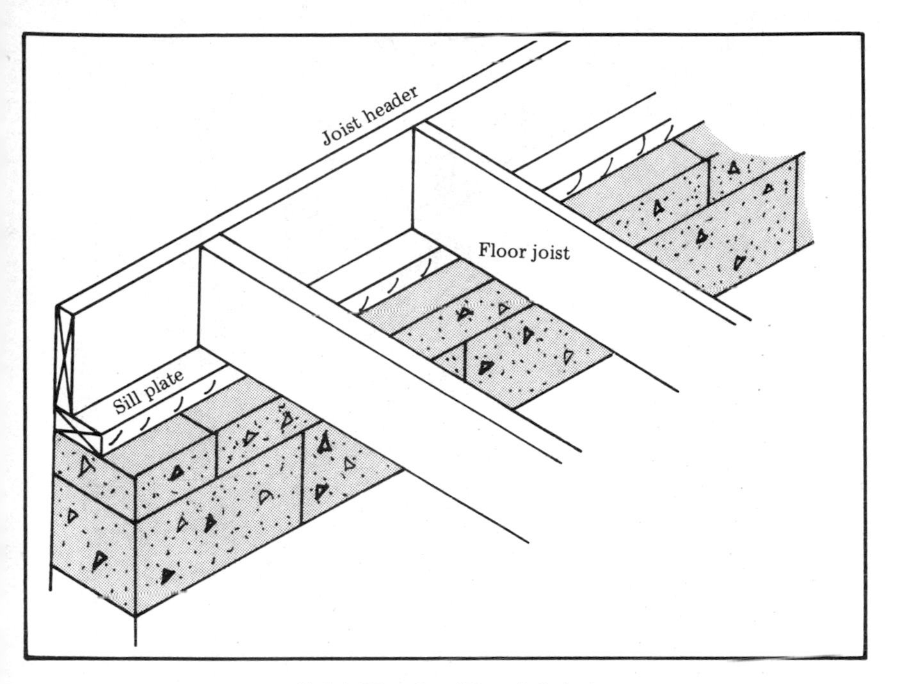

Joist Header (Band Joist)
Figure 6-7

2. *Front section:*

40.0' x .75 =	30
Starter	1
Double joists (3)	3
	34

Total: 34 — 2" x 10" x 16'

Note: No deductions were made for the stair opening. The cut-outs will be used for headers.

The total number of floor joists required in Figures 6-5 and 6-6 is:

1. *First Floor:*
   23 — 2" x 10" x 16' (613 bd. ft.)
   14 — Floor trusses
2. *Second Floor:*
   34 — 2" x 10" x 16' (907 bd. ft.)
   36 — 2" x 10" x 12' (720 bd. ft.)

Headers

Joist headers (or band joists) are at the end of the joists (Figure 6-7). Joist headers prevent lateral movement of the floor joists and provide a nailing base for the subfloor.

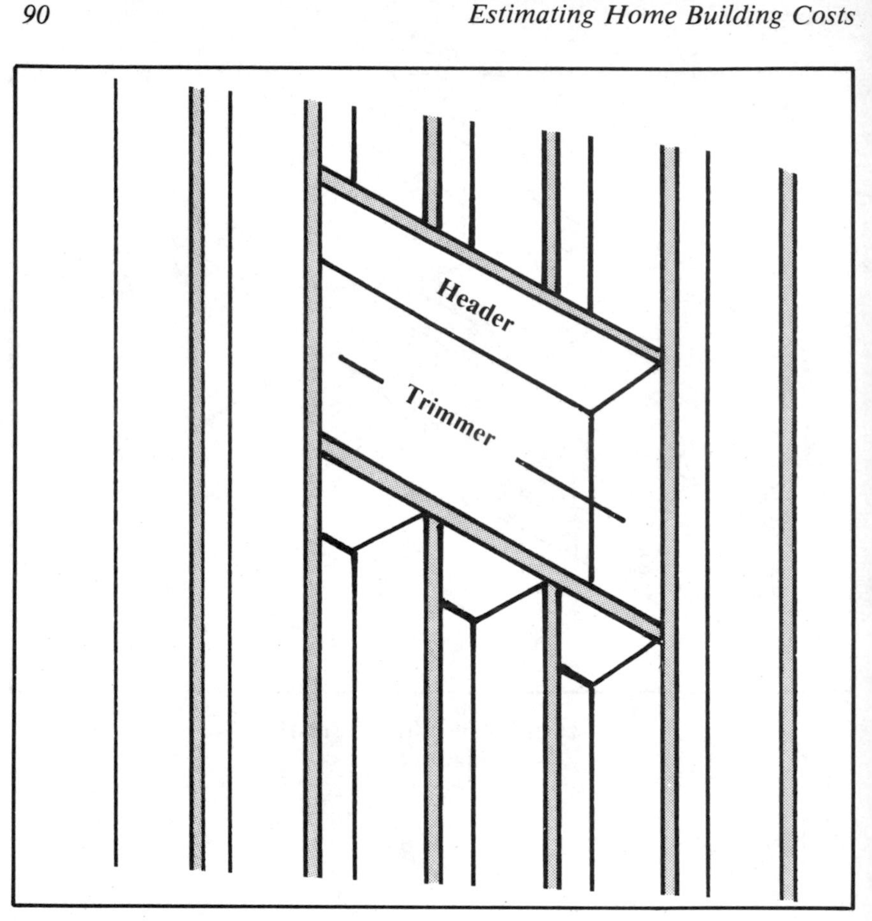

Joist Header and Trimmer
Figure 6-8

To estimate the quantity of joist headers (or band joists) required, divide the total lineal feet of the headers by the length of the framing to be used.

Example: How many pieces of 2'' x 10'' x 16' (the desired length for the headers) will be required for Figures 6-5 and 6-6?

Solution:

Lineal feet of joist headers

First floor	26'-0''
Second floor (40' + 40')	80'-0''
(*) Headers for stair openings	-0-
	106'-0''

106.0' divided by 16.0' = 6.63 (or 7)
Total: 7 — 2'' x 10'' x 16'

(*) The cut-outs from the joists will be used for these headers.

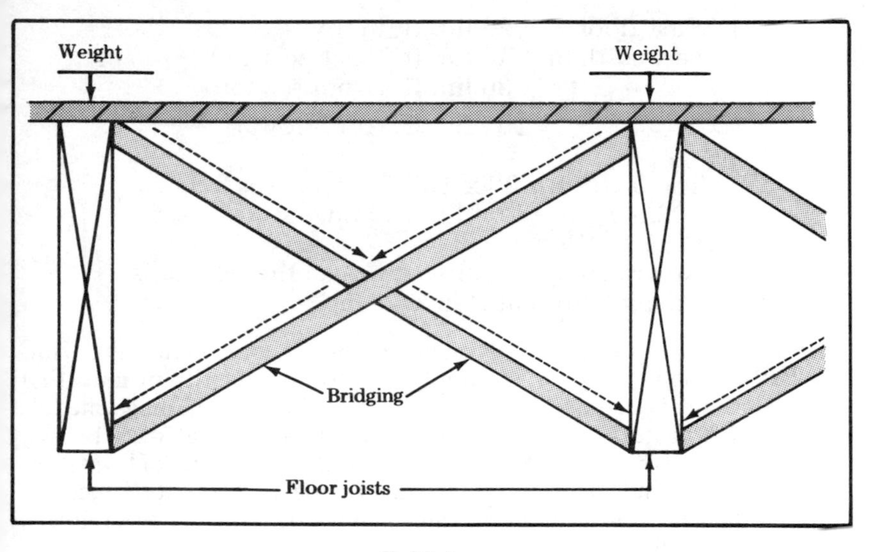

Bridging
Figure 6-9

Trimmers
The joist adjacent to an opening into which the header is framed is called a trimmer, (Figure 6-8). Most trimmers are estimated as floor joists.

Bridging
Bridging should be placed in spans over 8 feet between floor joists to stiffen them against twisting and to transfer the floor load from one joist to the adjacent joists (Figure 6-9). Some of the different types of bridging are:

1. *Wood cross bridging:* (Figure 6-9). When this type of bridging is used, 1'' x 3'' or 1'' x 4'' is recommended. There should be two 7d or 8d nails at each end. Estimate the quantity of bridging by adding the lineal feet in each row. Multiply by 2 if 2'' x 8'' or 2'' x 10'' joists are used. For 2'' x 12'' joists, multiply by 2.5 for the total lineal feet of bridging required. Allow 3.5 pounds of 8d nails per 100 lineal feet of bridging.

2. *Metal type bridging:* Order metal bridging according to the joist size and spacing. Estimate two pieces of metal bridging for each space between joists. For joists @ 16'' o.c. multiply the lineal feet of each run for the bridging by .75 (75%). For 24'' o.c. multiply the lineal feet of each run by .50 (50%).

3. *Solid bridging:* Use the same size members as the floor joists. They should be tightly fitted between joists.

The bridging in Figures 6-5 and 6-6 will be 1'' x 3'' wood cross bridging. The lineal feet required will be:

(*) First floor 26 lin. ft. in row
Second floor 40 lin. ft. (back section)
(**) 36 lin. ft. (front section)
102 lin. ft. (for 3 rows)

The factor for 2" x 10" floor joists is 2.0
102.0' x 2.0 = 204 lin. ft. bridging
Order: 210 lin. ft. 1" x 3"
(*) No bridging is required between the floor trusses
(**) Bridging stops at stair opening

Estimating the quantity of nails for residential construction is a "hit-or-miss" estimate many times. In addition to the quantity of nails that would normally be required following the recommended nailing schedule there is much waste. Extra nails are used for temporary scaffolds, braces, etc. Few factor tables are 100% correct, but the following factors allow for 15% waste and extras and should provide a reasonably accurate guideline to follow in estimating the quantity of nails for floor joists, headers and bridging.

16d common nails (approximate 54 nails per lb.)
Allow 10 lbs. per 1000 bd. ft.

Example: $\frac{2240 \text{ bd. ft. x } 10}{1000} = 22.40 \text{ lbs.}$

8d common nails (approximate 110 nails per lb.)

1. Toenail joist header to sill plate, and toenail floor joists to girder and plate:

Allow 2.5 lbs. per 1000 bd. ft.

Example: $\frac{2427 \text{ bd. ft. x } 2.5}{1000} = 6.07 \text{ lbs.}$

2. Bridging (four nails per pieces)
(*) Allow 3.5 lbs. per 110 lin. ft.

Example: $\frac{210 \text{ lin. ft. x } 3.5}{100} = 7.35 \text{ lbs.}$

(*) If 7d box nails are used, allow 2 lbs. per 100 lin. ft. This includes an allowance for waste and extras.

The material list for floor joists, headers, trimmers, bridging and nails in Figures 6-5 and 6-6 is:

Floor joists	57 — 2" x 10" x 16' (1520 bd. ft.)
	36 — 2" x 10" x 12' (720 bd. ft.)
Floor trusses	14 (designed by factory)
Headers	7 — 2" x 10" x 16' (187 bd. ft.)
Trimmers	0 — (estimated in floor joists)
Bridging	210 lin. ft. 1" x 3"
Nails	24 lbs. 16d common (including trusses)
	14 lbs. 8d common

(*) Enter on Line 6.2 Cost Estimate Sheet

(*) If the house is factory built and any of the above material is included in the package price of the house do not enter it on Line 6.2. It will be included in the house package price in the next chapter.

Subfloor

The subfloor, some times called the floor deck or floor sheathing, is applied over the floor joists and is the base for the finish foor. The material for subfloors in residential construction is usually wood boards or plywood, but plywood is more common.

Wood boards

The minimum thickness for wood boards is ⅜ inch and the maximum width is 8 inches. The boards may be installed diagonal to, or at right angles to floor joists. The end cut should be parallel to and over center of joists. Maximum joist spacing should be 16 inches o.c. except when 25/32 inch strip flooring is installed at right angles to joists and subfloor is installed diagonally. In this case, spacing may be 24 inches o.c. The wood boards should be nailed to joists at each bearing with 8d common or 7d threaded (anchor-down) nails. Use two nails in 6 inch boards and three in 8 inch boards.

Waste Factors
(Multiply area by factor for quantity)

Nominal Size	Laid at Right Angles to Joists	Laid Diagonal To Joists
1 x 6 S4S	1.14	1.20
1 x 8 S4S	1.12	1.17
1 x 6 T&G	1.20	1.25
1 x 8 T&G	1.15	1.20

7d threaded nails (approximate 172 nails per lb.)
(*) Allow: 27 lbs. per 1000 square feet.

8d common nails (approximate 111 nails per lb.)
(*) Allow: 40 lbs. per 1000 square feet.
(*) Includes 15% allowance for waste and extras

Example: If 1'' x 8'' S4S boards are to be used for the subfloor and are to be laid diagonal, how many board feet (one square foot equals one board foot for one inch boards) will be required for the floor system that is 40' x 26'? How many lbs. 7d threaded nails will be required allowing for waste and extras?

Answer: 40' x 26' = 1040 sq. ft.

1040 sq. ft. x 1.17 (factor) = 1216.80 or *1217 bd. ft.*

$$\frac{27 \text{ lbs.} \times 1040 \text{ sq. ft.}}{1000} = 28.08 \text{ or } \textit{28 lbs. nails}$$

The waste factors listed for lumber and nails are only guidelines. There is no substitute for experience in estimating material.

Example: If 1410 board feet of 1'' x 8'' S4S were used to install the subfloor for 42' x 28' (1176 square feet) the factor would be:

$$\frac{1410 \text{ bd. ft.}}{1176 \text{ sq. ft.}} = \textit{1.20 factor}$$

If 29 pounds 7d threaded nails were used for the 1176 square feet of subfloor the nail allowance would be:

(*) Proportion

$$29 : x = 1176 : 1000$$
$$x = 24.66 \text{ or } \textit{25 lbs. per 1000 bd. ft.}$$

(*) A detailed explanation on proportions and how they can be used in estimating is in Labor Costs in this chapter.

Plywood

Material shall be Structural-Interior type or Exterior type. Exterior type should be used when any surface or edge is exposed to weather. Plywood should be installed with the face grain at right angles to the joists and staggered so end joints break over different joists in adjacent panels.

The limiting factor in plywood subfloors is deflection under loads at the edges. Plywood must be continuous over two or more joists with the face grain across supports. Nail the plywood to joists at each bearing with 8d common or 6d threaded nails spaced 6 inches o.c. along all panel edges and 10 inches o.c. along intermediate joists (Figure 6-10).

Plywood Thickness (Inch)	Maximum Span (Inch)	Nail Size
1/2	16	6d threaded or 8d common
5/8	16	8d common
5/8, 3/4	20	8d common
3/4, 7/8	24	8d common

6d threaded nails (approximate 225 nails per pound)
(*) Allow: 8 pounds per 1000 square feet.

8d common nails (approximate 111 nails per pound)
(*) Allow: 15 pounds per 1000 square feet (16 inch o.c.)
(*) Allow: 11.5 pounds per 1000 square feet (24 inch o.c.)
(*) Includes 15% allowance for waste and extras.

In estimating plywood, take the square feet of the floor area and divide by the square feet in one piece of plywood and round off to the next whole number. *Example:* if there is 1040 square feet of floor area and 4 x 8 (32 square feet) plywood is used, the number of pieces required will be:

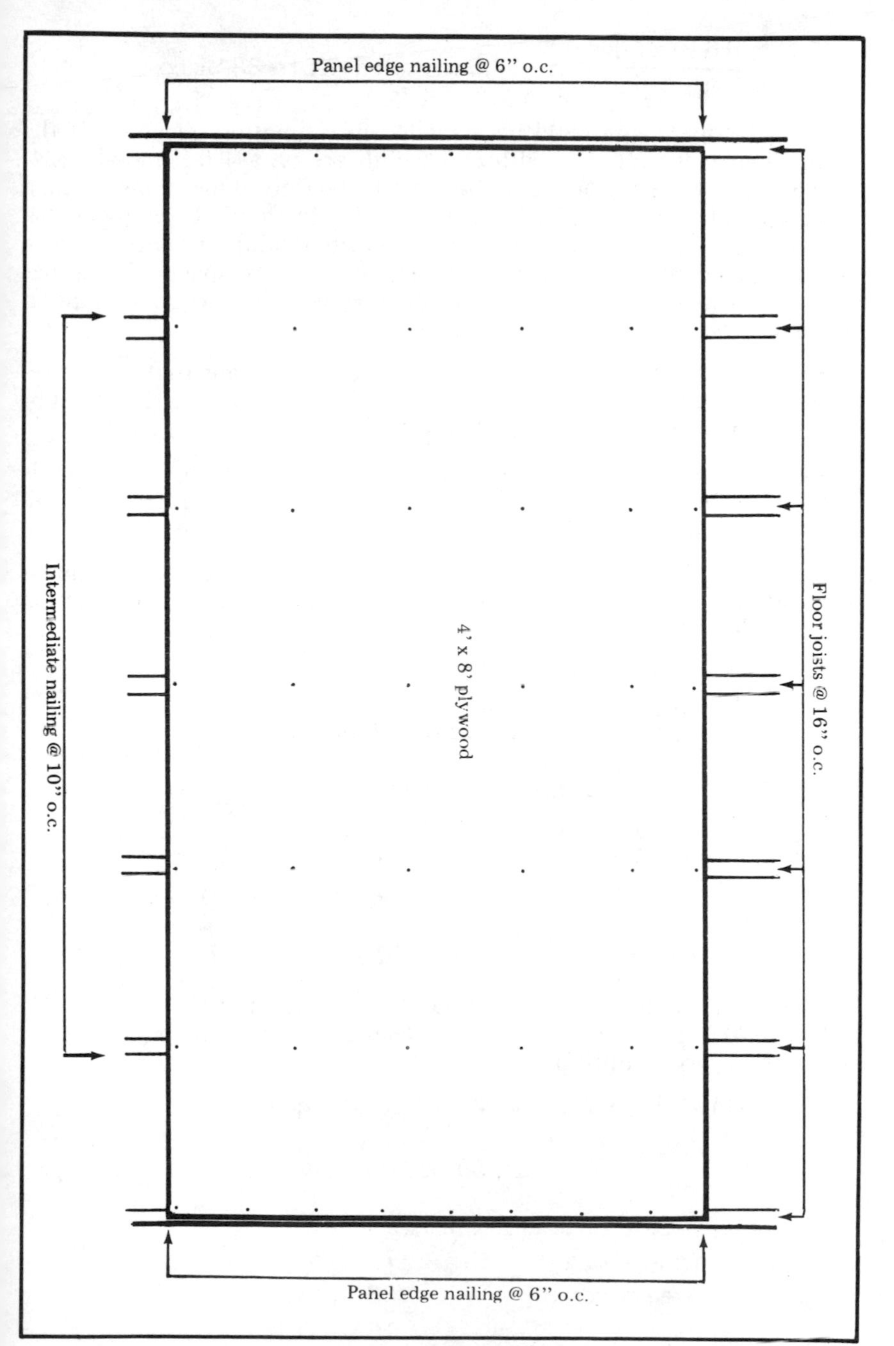

Figure 6-10
Plywood Nail Detail

$$\frac{1040 \text{ sq. ft.}}{32} = 32.5 \text{ rounded off to 33 pieces}$$

Only large openings should be deducted in estimating plywood. If the floor dimensions for the width and length are on a 4-feet module (24' 48', 64') the waste factor should be zero. If the floor dimensions are on a 2-feet module (26', 34', 50', 66') the waste factor should be between 3% - 5% If any of the dimensions vary from the module system the waste factor can be much higher (Figure 6-11). Uneven joist spacing is another factor that adds to waste. A plywood layout drawn to scale helps in estimating waste (Figures 6-11 and 6-12).

Example: Estimate the quantity of plywood and nails that will be required for the first and second floors in Figures 6-11 and 6-12. The subfoor will be 4' x 8' plywood nailed down with 8d common nails.

1. In Figure 6-11 the right side with the floor elevation of 106.14' is 24'-2¼'' x 26'-0''. The floor trusses are 24 inch o.c. and the plans call for ¾ inch T&G subfloor.

2. The 2¼ inches can be disregarded in the 24'-2¼'' because the dimension is to the outside of the foundation and the floor trusses set back on the outside wall.

24.0' x 26.0' = 624 sq. ft.

$$\frac{624 \text{ sq. ft.}}{32} = 19.50 \text{ or } \textit{20 pieces}$$

Calculations for Waste

20 pcs. @ 32' = 640 sq. ft.
19.5 pcs. @ 32' = 624 sq. ft.
16 sq. ft. waste (.03%)

Order 20 — ¾'' x 4' x 8' T&G plywood

3. The left side in Figure 6-11 with the floor elevation of 104.90' is 15'-5'' ¾'' x 26'-0''. The floor joists are on 16 inch o.c. and the plans call for 5/8 inch plywood subfloor.

4. 15.48' (15'-5¾'') x 26.0' = 402.48 or 403 square feet.

$$\frac{403 \text{ sq. ft.}}{32} = 12.59 \text{ or } \textit{13 pieces}$$

Calculations For Waste

13 pcs. @ 32'	=	416 sq. ft.
Area covered	=	403 sq. ft.
		13 sq. ft.
Stair cut out	=	28 sq. ft.
		41 sq. ft. waste (10%)

Order 13 - 5/8'' x 4' x 8' C-D plywood

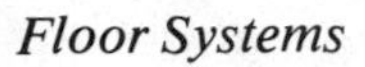

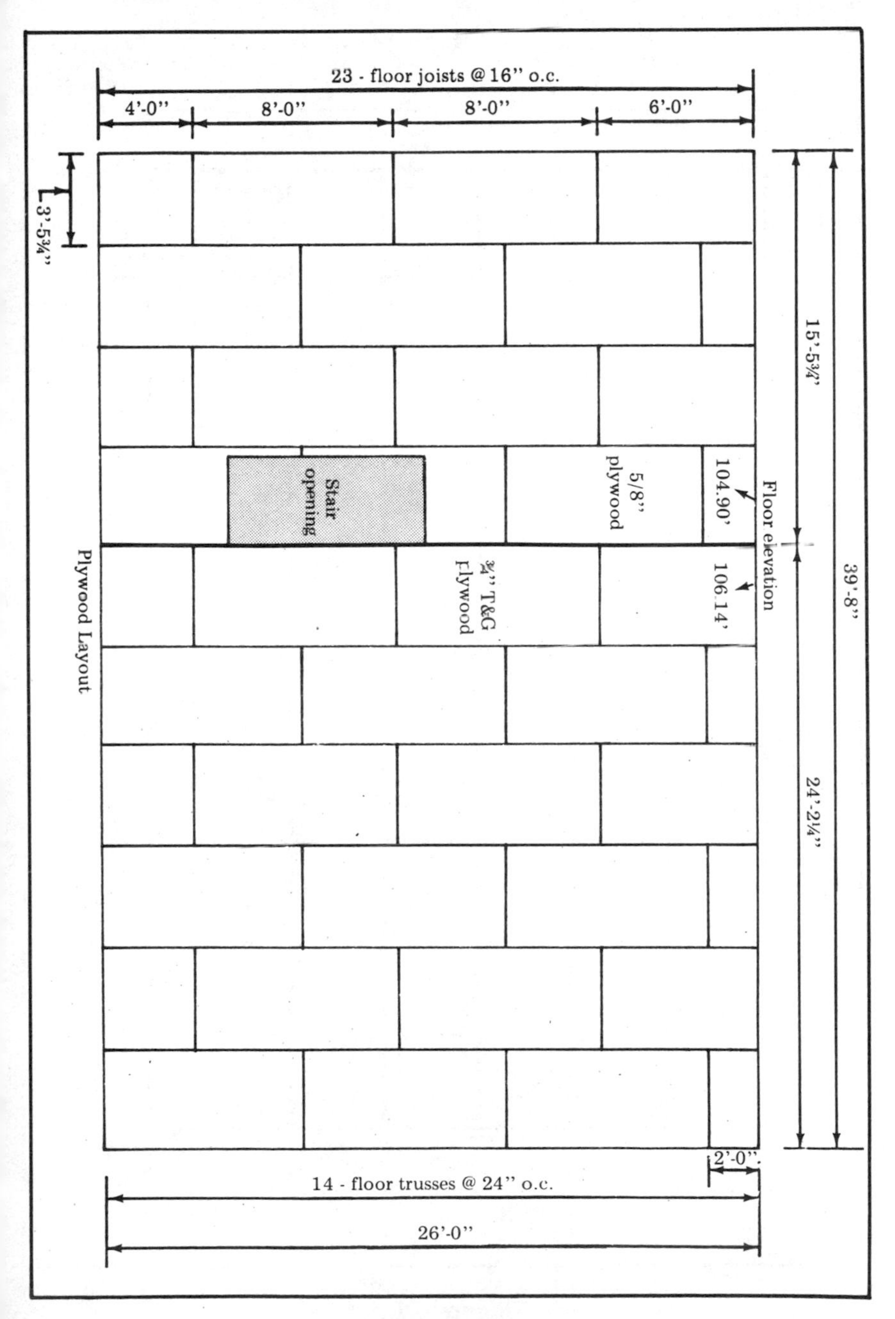

Figure 6-11
Plywood Layout
First Floor

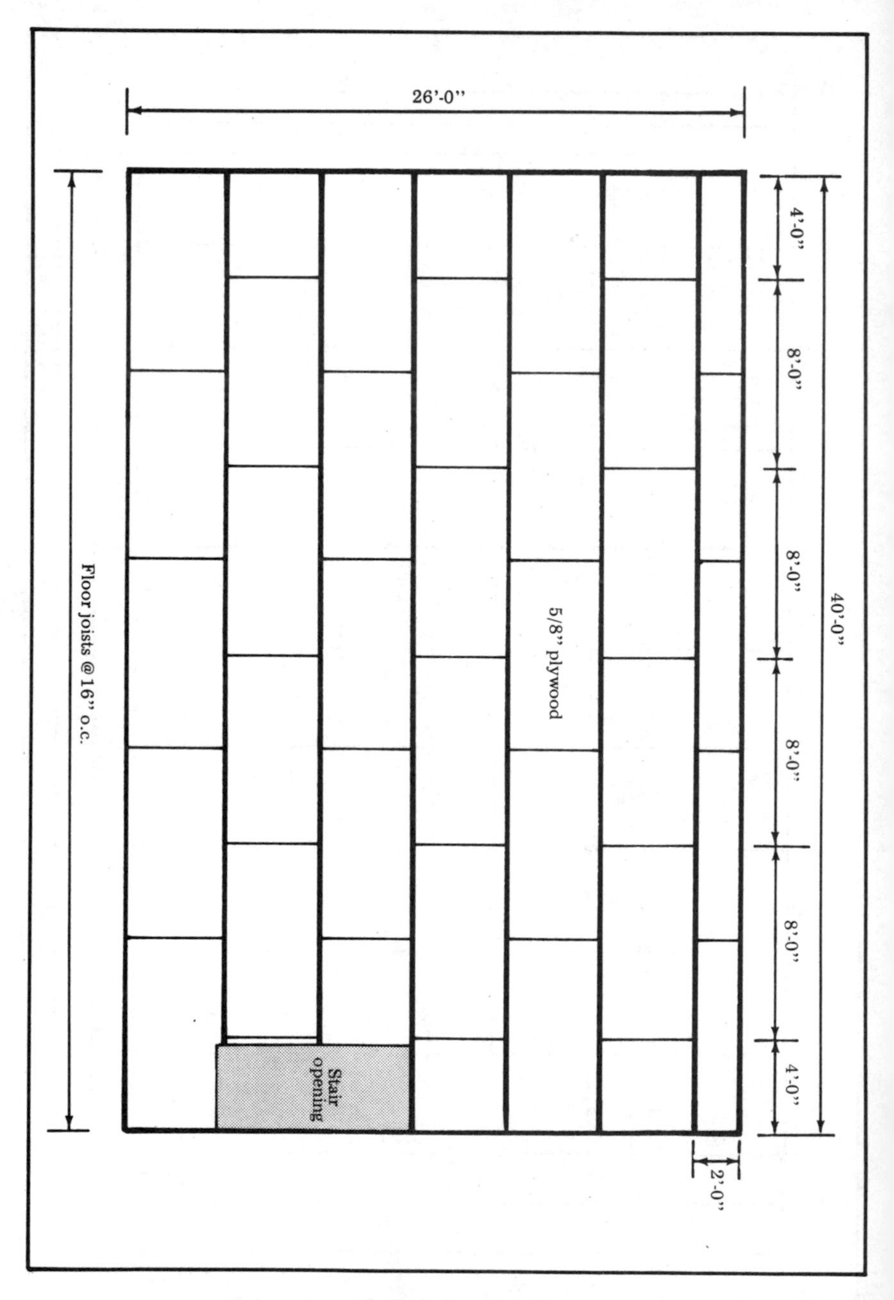

Figure 6-12
Plywood Layout
Second Floor

5. In Figure 6-12 (second floor) the dimensions are 40'-0'' x 26'-0''. The floor joists are 16 inch o.c. and the plans call for 5/8'' plywood subfloor.

40.0' x 26.0' = 1040 sq. ft.

$$\frac{1040 \text{ sq. ft.}}{32} = 32.50 \text{ or } \textit{33 pieces}$$

Calculations For Waste

33 pcs. @ 32'	=	1056 sq. ft.
32.5 pcs. @ 32'	=	1040 sq. ft.
		16 sq. ft.
Stair cut out	=	28 sq. ft.
		44 sq. ft. waste (.04%)

Order 33 — 5/8'' x 4' x 8' C-D plywood

The quantity of 8d common nails required will be:

First floor: 624 sq. ft. (joists 24 inch o.c.)

$$\frac{624 \times 11.5 \text{ (factor)}}{1000} = 7.18 \text{ lbs.}$$

First floor: 403 sq. ft. (joists 16 inch o.c.)

$$\frac{403 \times 15 \text{ (factor)}}{1000} = 6.05 \text{ lbs.}$$

Second floor: 1040 sq. ft. (16 inch o.c.)

$$\frac{1040 \times 15 \text{ (factor)}}{1000} = 15.60 \text{ lbs.}$$

28.83 or *29 lbs.*

Order 29 lbs. 8d common nails

Order the following for the subfloors in Figures 6-11 and 6-12.

5/8'' C-D plywood

First floor		13 pieces
Second floor		33 pieces
	Total	46 pieces

46 - 5/8'' x 4' x 8' C-D plywood

20 - 3/4'' x 4' x 8' T&G plywood

29 - lbs. 8d common nails

(*) Enter on Line 6.3 Cost Estimate Work Sheet

(*) If the house is factory built and the above material is included in the package price of the house do not enter it on Line 6.3, it will be included in the package price in the next chapter.

Labor Costs for Floor Systems

Labor costs are always only estimates and are difficult to calculate. There are many labor factor tables available for estimating the manhours for each phase of residential construction, and they are good, but the most accurate estimates are from past experience on similar jobs. The use of proportions can save the builder or estimator much time in estimating, especially in manhours and materials where there is a high waste factor such as masonry cement and nails. A brief review of proportions and how they can help the estimator follows.

Proportions

The radio of two numbers is the quotient of one quantity divided by another of the same kind, or the relationship of one number to another. This relationship is indicated by the sign (:). Example: 6:8 is the ratio of 6 to 8 and is equivalent to the fraction obtained by dividing the first number by the second. Thus, ¾ is the ratio of 6 to 8.

When two ratios are equal it is called a proportion. Thus, 6:8 = 12:16. The first and last numbers (or end numbers) are called "extremes" and the second and third numbers (or middle numbers) are called "means". In this ratio the "extremes" are 6 and 16 and the "means" are 8 and 12.

The product of the "means" equals the product of the "extremes". *Example:*

$$6 : 8 = 12 : 16$$

"Means" $8 \times 12 = 96$

"Extremes" $6 \times 16 = 96$

If any three terms of the proportion are known the fourth can be found. The product of two divided by the third known number produces the fourth number. The unknown number in a proportion is usually designated as (x). *Examples:*

$$6 : x = 12 : 16$$

Solution: Product of "extremes" $6 \times 16 = 96$

(known number) $\frac{96}{12} = 8$

$$x = 8$$

$$6 : 8 = 12 : x$$

Product of "means" $8 \times 12 = 96$

(known number) $\frac{96}{6} = 16$

$$x = 16$$

The solution in a proportion is easier to understand when numbers representing the same things are in the same ratio such as, dollars to dollars, man-hours to man-hours, board feet to board feet, pounds to pounds, and so on.

Example: If recent payroll records show that 161 man-hours were required to install the floor system for the first floor of a residential house with 2224 square feet and a basement, how many man-hours should be estimated for the floor system in a house with 1084 square feet and a basement?

Solution:

$$\underbrace{161 : x}_{\text{Man-Hours}} = \underbrace{2224:1084}_{\text{Square Feet}}$$

Known Ratio

Product of "extremes" (end numbers) $161 \times 1084 = 174524$

$$\text{(known number)} \quad \frac{174524}{2224} = 78.47$$

$$x = 78.47 \text{ or } \textit{79 man-hours}$$

Example: If 64 bags of masonry cement were used to lay 2139 masonry blocks, how many bags should be estimated for 1784 blocks?

$$64:x = 2139:1784$$

Product of "extremes" $64 \times 1784 = 114176$

$$\frac{114176}{2139} = 53.38$$

$$x = 53.38 \text{ or } \textit{54 bags}$$

Example: If 29 pounds 7d threaded nails were used to install 1176 square feet 1'' x 8'' S4S subfloor, what nail factor should be used per 1000 square feet for another job using the same type of material?

$$29:x = 1176:1000$$

Product of "extremes" $29 \times 1000 = 29000$

$$\frac{29000}{1176} = 24.66$$

$$x = 24.66 \text{ or } \textit{25 pounds}$$

The nail factor will be 25 pounds per 1000 square feet

In estimating the labor to install the floor system for the first and second floors in Figures 6-11 and 6-12, reference was made to the payroll records on a recent job for a residential house with a basement and using the same type of materials. From these records we saw that 161 man-hours were require for 2224 square feet on the first floor and 118 man-hours were required for 1360 square feet on the second floor. Figure 6-11 shows a first floor of 1032 square feet. The estimated number of man-hours to install this floor system will be:

$$161:x = 2224:1032$$

$$x = 74.71 \text{ or } \textit{75 man-hours}$$

The second floor has 1040 square feet, so the estimated number of man-hours to install it will be:

$$118:x = 1360:1040$$

$$x = 90.24 \text{ or } \textit{91 man-hours}$$

The total man-hours to install both floor systems is estimated at:

First floor	75
Second floor	91
Total	166 man-hours

The workmen who will be assigned to the job will be one foreman at $8.50 per hour, two carpenter helpers at $6.25 per hour and three laborers at $5.00 per hour.

$$\frac{(\text{man-hours})\ 166}{(\text{workmen})\ 6} = 27.67 \text{ or } \textit{28 man-hours each}$$

28 hours @ $8.50 x 1	=	$238.00
28 hours @ 6.25 x 2	=	350.00
28 hours @ 5.00 x 3	=	420.00
(*) Labor cost		$1008.00

(*) Payroll taxes and insurance will be added later.

(Enter labor cost on Line 6.4 Cost Estimate Work Sheet)

Cost Estimate Work Sheet For Floor System

6.1 Sill plate, girder, ledger or joist hangers:
_____ pcs. Sill plate (size _____) @ _____ = $_____
_____ pcs. Girder (size _____) @ _____ = _____
_____ pcs. Girder (size _____) @ _____ = _____
_____ pcs. Ledger (size _____) @ _____ = _____
_____ pcs. Joist hangers (size _____ @_____ = _____
_____ lbs. Nails (size_____) @ _____ = _____
_____ lbs. Nails (size _____) @ _____ = _____
$_____ $_______

6.2 Floor joists, headers, trimmers and bridging:
_____ pcs. Floor joists (size _____) @ $_____ = $_____
_____ pcs. Floor joists (size _____) @ _____ = _____
_____ pcs. Floor joists (size _____) @ _____ = _____
_____ pcs. Floor joists (size _____) @ _____ = _____
_____ pcs. Floor trusses (size_____) @_____ = _____
_____ pcs. Headers (size _____) @ _____ = _____
_____ pcs. Headers (size _____) @ _____ = _____
_____ pcs. Trimmers (size _____) @ _____ = _____
_____ lin. ft. Bridging (size_____) @_____ = _____
_____ lbs. Nails (size _____) @ _____ = _____
_____ lbs. Nails (size _____) @ _____ = _____
$_____ _______

6.3 Subfloors:
_____ bd. ft. Wood boards (size_____) @_____ = $_____
_____ pcs. Plywood (size _____) @ _____ = _____
_____ pcs. Plywood (size _____) @ _____ = _____
_____ lbs. Nails (size _____) @ _____ = _____
_____ lbs. Nails (size _____) @ _____ = _____
$_____ $_____

Total for page 1 (brought forward) $______

Cost of material $________

Sales tax (____%) ________

Total cost of material $________(1)

6.4 Labor costs for floor system: $________(2)

Cost of floor system (add lines (1) and (2) $________

(Enter on Line 6 - Form 100)

Chapter 7
Superstructure

The superstructure of the house as defined in this chapter consists of the following:

1. Framing for exterior walls
2. Framing for interior walls
3. Sheathing for exterior walls
4. Framing for roof system
5. Sheathing for roof
6. Stringers for stairs from first floor to second floor
7. Framing and sheathing for porches
8. Labor and other materials necessary to assemble the superstructure

If the house is factory-built, or a sectional house, the component parts will include all of the above material fabricated in sections plus the windows installed, prehung doors, exterior and interior trim and many optional materials to be explained later. Many houses, called Modular Houses, are completely finished at the factory, transported to the job site in sections and erected on the builder's foundation.

Exterior and Interior Walls

Sole and Top Plates

The sole, or bottom plate, is a horizontal member, rests on the subfloor and has the studs nailed to it (Figures 7-3 and 7-4). The plates in normal residential construction are 2 x 4s. The sole plate should be nailed to the floor joists with 16d common nails, or 3½ inch spiral thread nails spaced not more than 16 inches o.c.

The top plates should be doubled for all bearing walls (Figures 7-3 and 7-4), and may be single for non-bearing walls as shown in Figure 7-5. However, the extra cost of the double plate for a non-bearing wall is often offset by the added cost of handling and cutting the studs, and by the interruption in the continuity of the work. For this reason many builders use double top plates on all walls.

Ends of top plates on lower members should occur over studs. Joints in upper member of plates should occur at least 24 inches from joints in lower members. Use two 16d common nails at each stud for the lower top plate member, and nail the upper plate to the lower plate with 16d common or 3½ inch spiral thread nails 16 inches o.c.

When estimating the sole and top plates, take-off the lineal feet of the exterior and interior walls on each floor separately. Some walls may differ from the normal 2 x 4 walls. For example, a wall may be 6 inches wide to receive the plumbing for a bathroom, while a closet partition may be only 2 inches wide to gain additional space. The estimator should take care not to overlook these different size walls. Door openings are not normally deducted unless they exceed 8 feet, such as a garage door. The cutouts from these openings will be used in many places during the construction of the house. If double top plates are to be used on all walls (as in this estimate), multiply the total lineal feet of the exterior and interior walls by the number of plates (in this estimate the number will be 3) to get the total lineal feet of top and sole plates required. If 2 x 4 x 12s are to be used, divide the total lineal feet of the plates by 12 to get the number of 2 x 4 x 12s that will be required. If 2 x 4 x 14s are used, divide by 14, and so on. Example: estimate the number of 2 x 4 x 12s that will be required for the sole and top plates in Figures 7-1 and 7-2. Double top plates will be used on all walls. There are no walls that deviate from the normal 2 x 4 walls.

Solution: **First Floor (Figure 7-1)**

Exterior walls ....192 lin. ft.
Interior walls...213 lin. ft.

Solution: **Second Floor (Figure 7-2)**

Exterior walls ....132 lin. ft.
Interior walls...185 lin. ft.

Total 722 lin. ft.

722 x 3 = 2166 lin. ft. plates

Less. . . 18 lin. ft. (garage doors)

2148 lin. ft.

$$\frac{2148}{12} = 179 \text{ (number of pieces)}$$

Order: 179 — 2 x 4 x 12. . . . . . (1432 board feet)

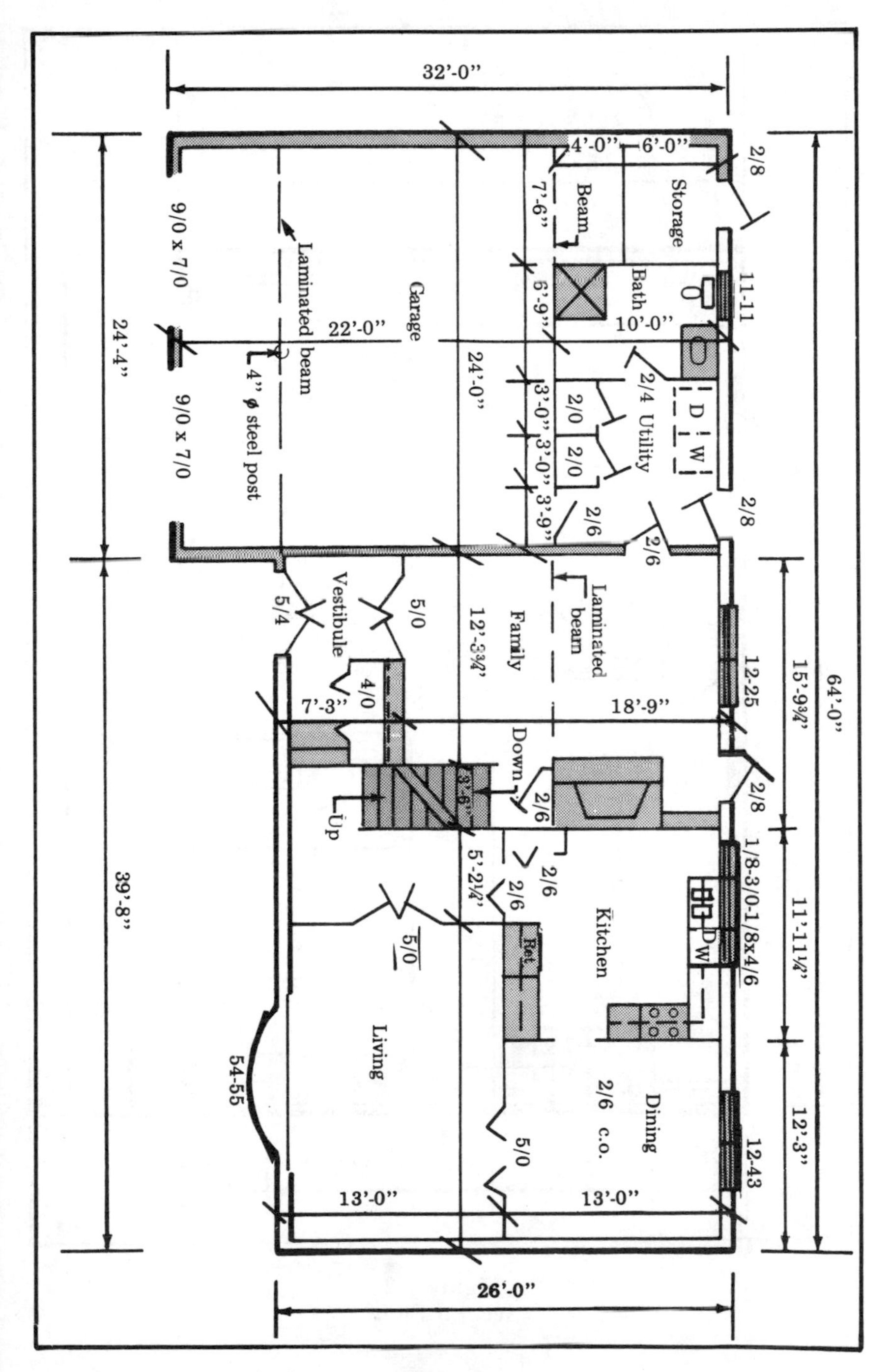

Figure 7-1
First Floor

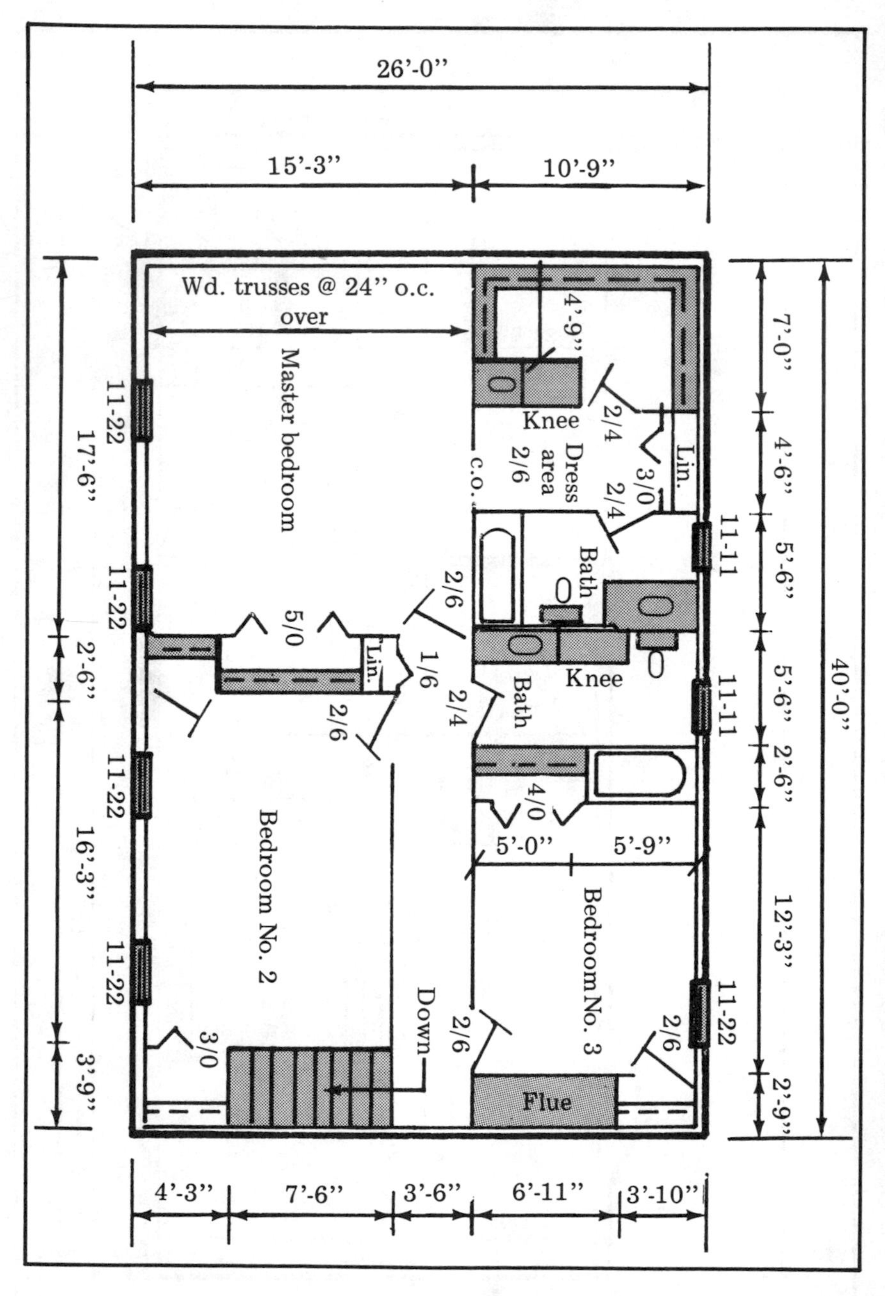

Figure 7-2
Second Floor

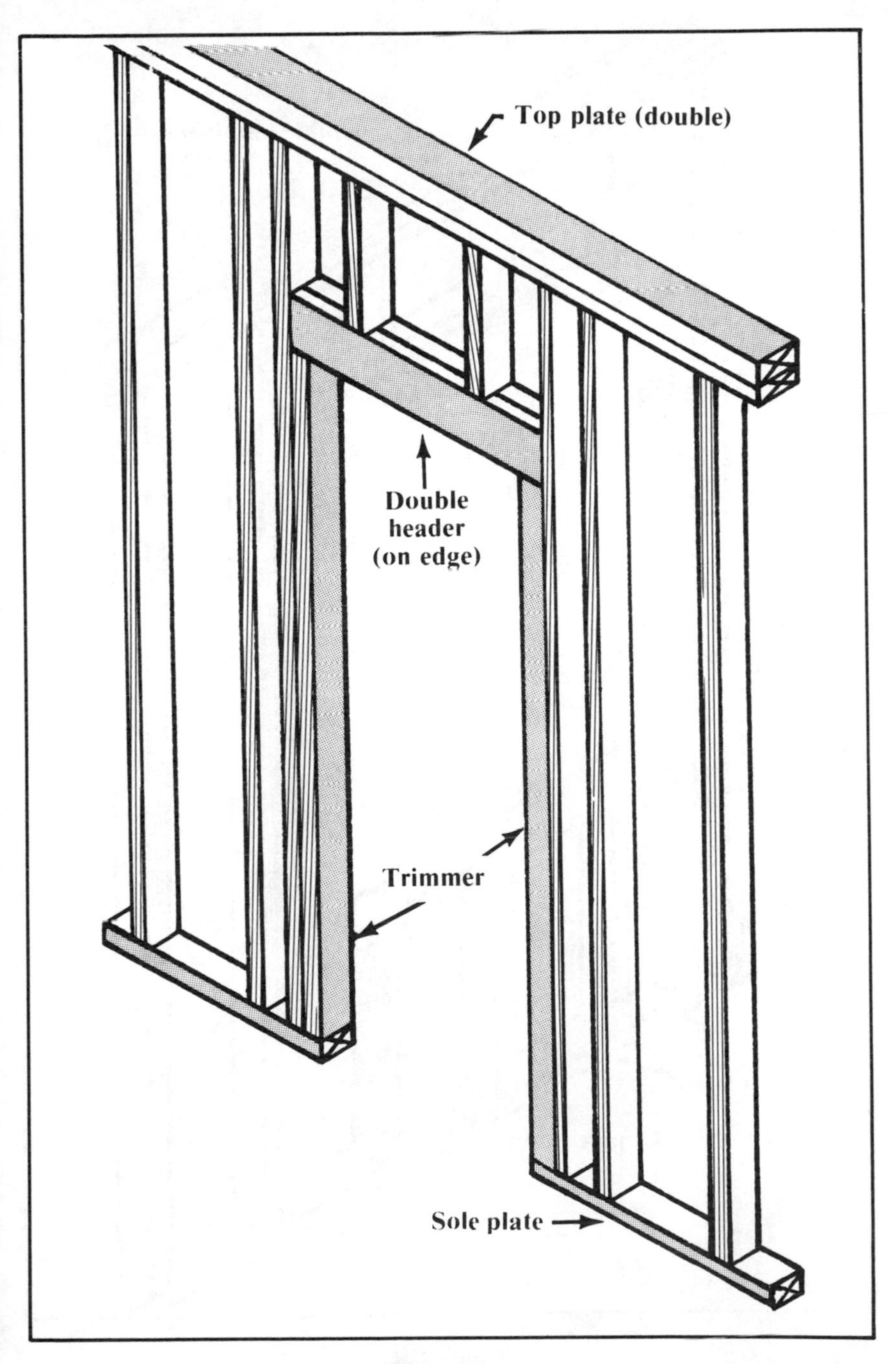

Figure 7-3
Bearing Wall

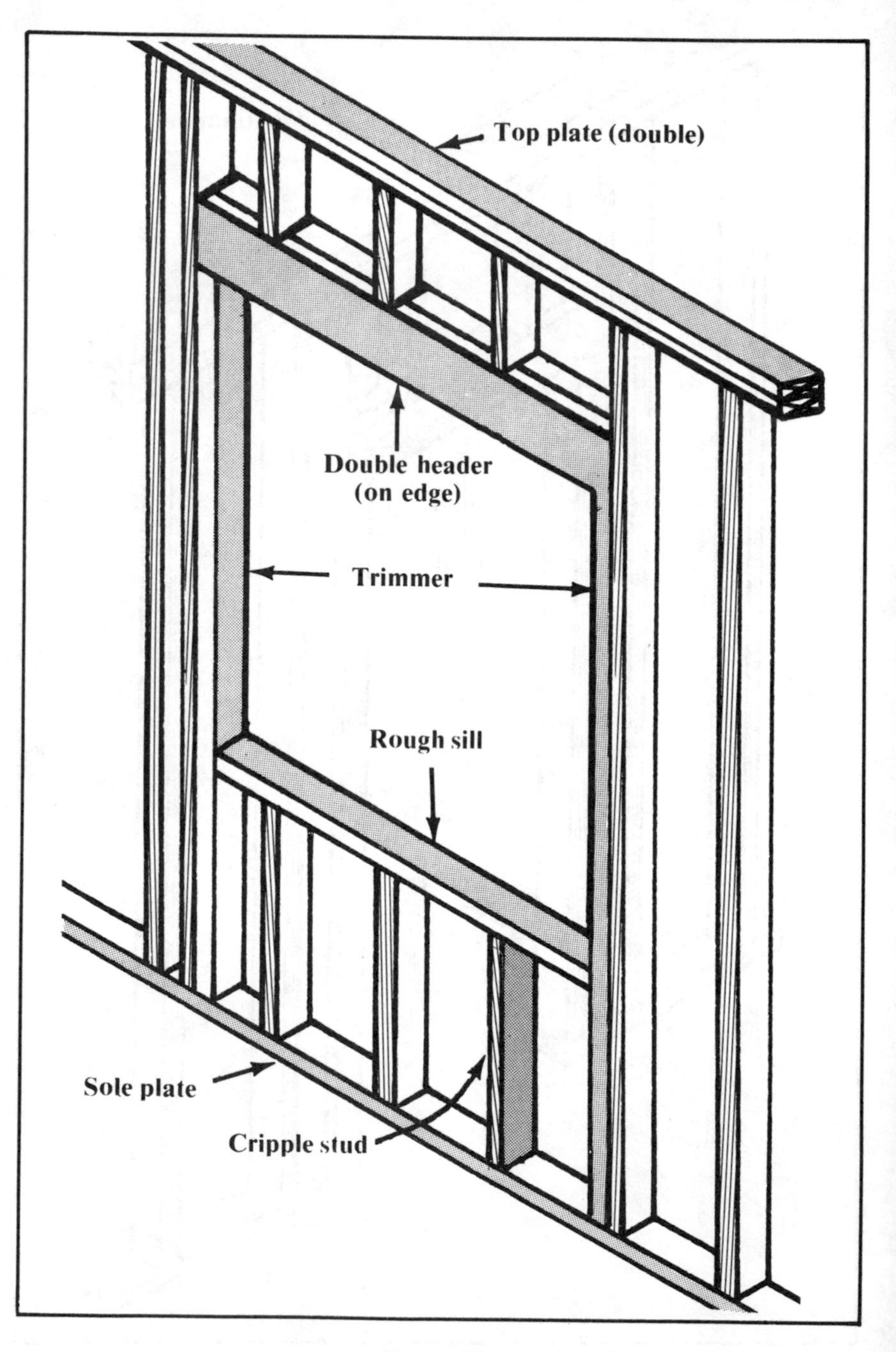

Figure 7-4
Bearing Wall

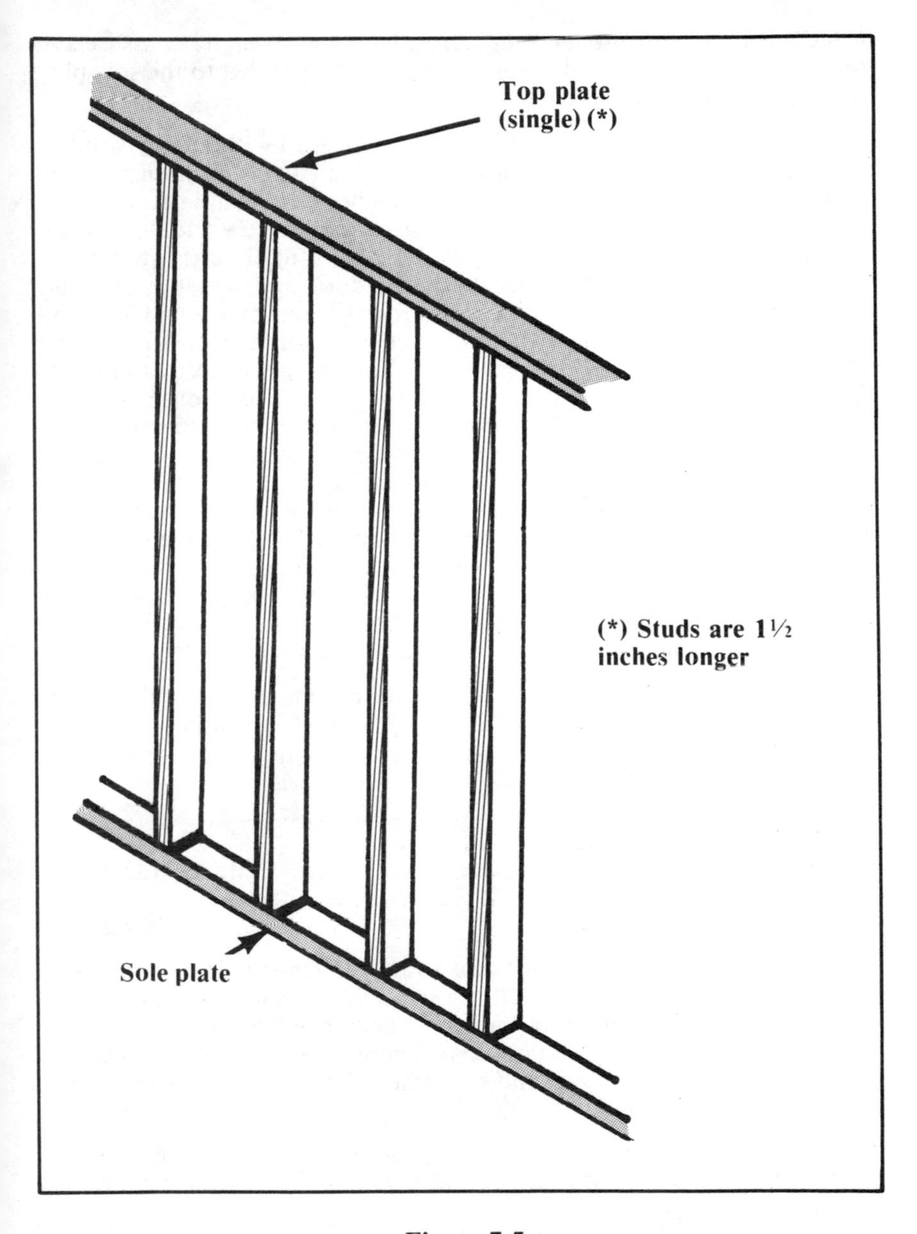

Figure 7-5
Non-bearing Wall

The studs are the vertical framing members in a wall to which the finish material is applied. They have been referred to as the bone structure of a house. The wall section of the blueprints will give the size, spacing and wall height for the stud lengths. The studs must be continuous lengths without splicing, and they must have the strength to assure safe support of the design loads. The studs should be nailed to the sole plate with four 8d or two 16d common nails.

Estimating the number of studs in a straight wall without any window or door openings or intersecting walls is not difficult. Example: a wall 12'0" in length with the studs spaced 16 inches o.c. would take 10 studs (12' x .75 = 9 plus 1). However, because of the window and door openings (Figures 7-3 and 7-4), corner posts (Figure 7-6), T-posts at intersecting walls (Figure 7-7), trimmer studs for door and window openings (Figures 7-3 and 7-4), blocking for partitions (Figures 7-6 and 7-7) and the many other uses for 2 x 4s during the construction of a house, an allowance must be made for these studs. You can get a close estimate of the number of studs that will be required for all walls in a house by allowing one stud for each lineal foot of wall space. Some estimators will make an additional allowance of two studs for each corner or an arbitrary allowance of 50-100 additional studs, depending on the size of the house. There is a common saying among builders: "Estimate the number of 2 x 4s you think you need, add a truck load and you may have enough." This is an extreme example, but it illustrates a point: always allow for more studs than the number calculated. Previous experience is very helpful in making this additional allowance. Also, do not overlook double walls where they are used for plumbing and sound effect.

The number of studs required for the exterior and interior walls for the first floor (Figure 7-1) and second floor (Figure 7-2) will be:

1. From the previous calculations, when estimating the sole and top plates, the lineal feet of walls is:

First floor exterior walls	192
First floor interior walls	213
Second floor exterior walls	132
Second floor interior walls	185
Total	722 lin. ft.

2. Allowing one stud for each lineal foot, 722 studs will be estimated. Note: these studs can be ordered precut to 92 ⅝ inches (adding 4½ inches for the sole and two top plates will give a ceiling height of 8'1⅛" from the sub-floor to the ceiling joists). There is very little difference between the cost of precut studs and the regular 2 x 4 x 8, but there is a big saving in the labor costs.

If rafters are used in place of roof trusses and there are gables, studs must be estimated for the gable ends. These studs are estimated as follows:

1. Multiply the width of the gable by .75 for 16 inch o.c. or by .50 if 24 inch o.c. and add one stud for the starter to get the number of studs for each gable.

2. Scale the length of the longest stud from the elevation on the blueprints and divide by 2 for the average stud length.

3. Multiply the number of studs in each gable by the average stud length for the total lineal feet of the gable studs.

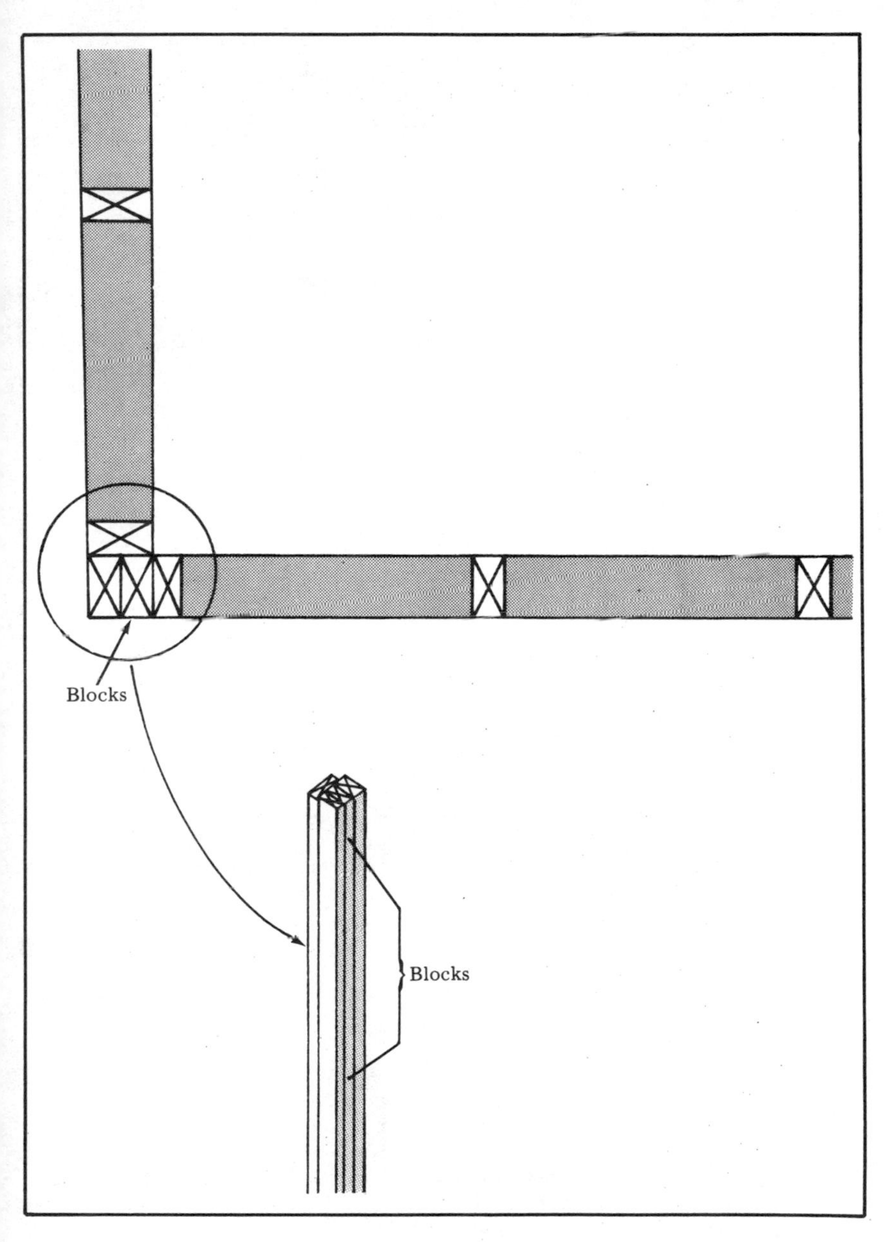

Figure 7-6
Corner Post

4. Divide the total lineal feet of the gable studs by 8 and round off to the next highest whole number for the number of 2 x 4 x 8 studs required for the gable.

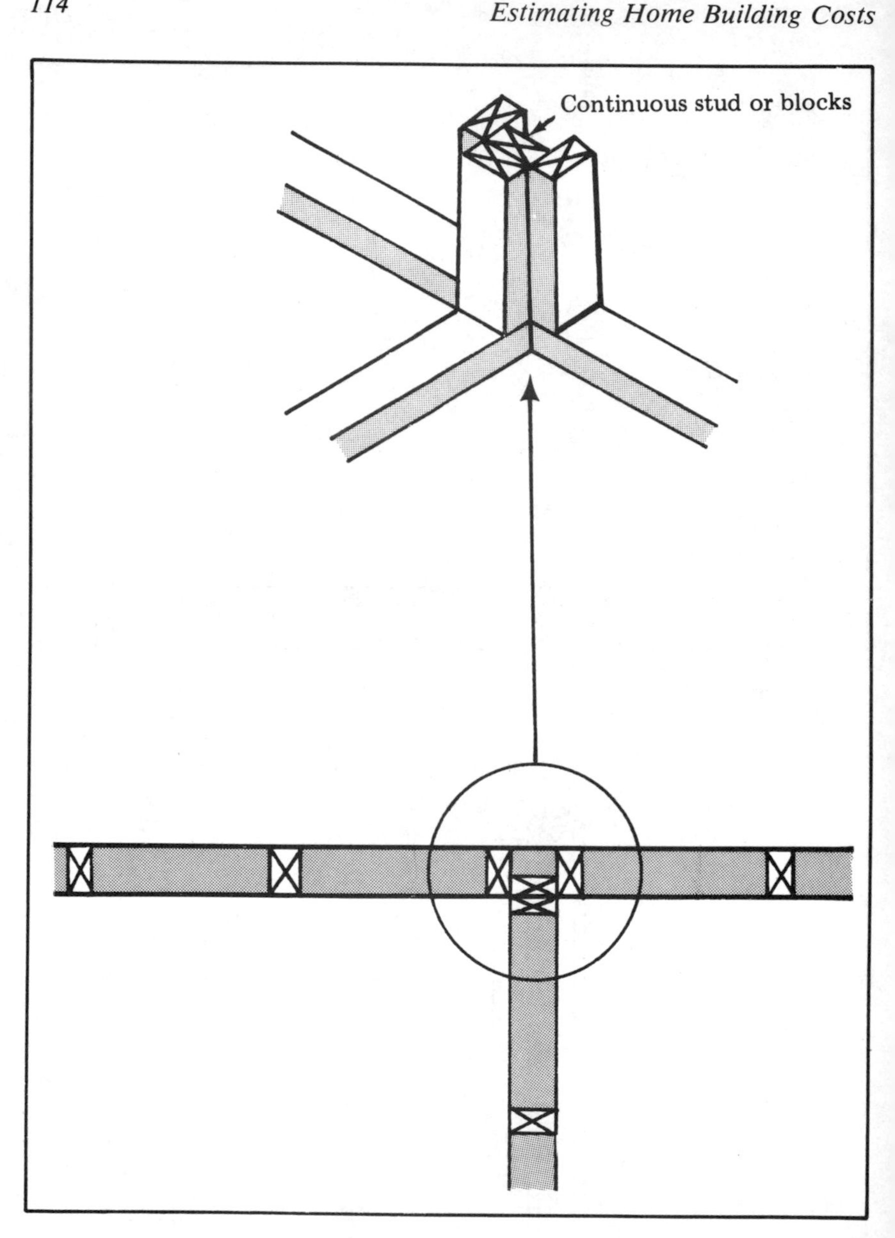

Figure 7-7
Wall Intersection

5. If all gables are the same size, as shown in Figure 7-10, multiply the number of studs in each gable by the number of gables for the total number of gable studs required. If gables vary in size, calculate each gable separately and total for the total number of studs required for all gables.

Example: From Figure 7-10 estimate the number of gable studs that will be required for the three gables plus the gables on the 6 foot offset for the garage area. The studs are spaced 16 inches o.c.

Solution:
1. The width of each main gable is 26'0''
26' x .75 = 19.50 or 20 studs plus 1 = 21 studs per gable

2. The longest gable stud is 1'6''. This length will be used for the average stud length because of the large gable vents.

3. 21 studs x 1.5' (1'6'') = 31.5 lin. ft.
31.5 lin. ft. x 3 gables = 94.5 lin. ft.
Add for two gables on offset = 11.0 lin. ft.
94.5 + 11.0 = 105.5 lin. ft.

$$\frac{105.5}{8} = 13.19 \text{ or } 14 \text{ studs}$$

Order: 14 - 2 x 4 x 8 studs for the gables

The total studs required for the exterior, interior and gable walls in Figures 7-1, 7-2 and 7-10 are:

Exterior and interior walls.	722 - 2 x 4 x 8
Gables	14 - 2 x 4 x 8
Allow for extras.	64 - 2 x 4 x 8
Total	800 - 2 x 4 x 8

Order: *800 - 2 x 4 x 8 studs. . . . . (4267 bd. ft.)*

Headers
Headers support the weight over openings (Figures 7-3 and 7-4). The size of the headers varies according to the span of the opening. The estimator must know the size including the length of each header and he must double the header length for the double header (Figures 7-8 and 7-9).

When estimating headers for interior doors, the length of each header is the door width plus 5 inches and is calculated as follows:

Thickness of door jambs (2 @¾'') . . .	1½''
Allowance for clearance	½''
Bearing on 2 studs (2 @1½'').	3 ''
Total	5 ''

Thus the header length for a 2'6'' door would be 2'11'' and because the header is doubled, it will require 5'10'' (2 x 2'11'') for both headers. One 2 x 4 x 8 will have to be estimated for this door opening (Figure 7-8).

Estimating the header lengths for the door and window openings in exterior walls requires more time and caution. For example, when windows

Door Width	Header Size (On Edge)	Header Length	Estimate for Each Double Header
2'-0"	2 - 2 x 4s	2'- 5"	1 - 2 x 4 x 8
2'-4"	2 - 2 x 4s	2'- 9"	1 - 2 x 4 x 8
2'-6"	2 - 2 x 4s	2'-11"	1 - 2 x 4 x 8
2'-8"	2 - 2 x 6s	3'- 1"	1 - 2 x 6 x 8
3'-0"	2 - 2 x 6s	3'- 5"	1 - 2 x 6 x 8
4'-0"	2 - 2 x 6s	4'- 5"	1 - 2 x 6 x 10
5'-0"	2 - 2 x 6s	5'- 5"	1 - 2 x 6 x 12
6'-0"	2 - 2 x 8s	6'- 5"	1 - 2 x 8 x 14

Headers - Interior Door Openings
Figure 7-8

Header Size (On Edge)	Maximum Width of Rough Stud Opening	Header Length
2 - 2 x 4s	3'-0"	Rough stud opening width plus bearing on two studs
2 - 2 x 6s	6'-0"	
2 - 2 x 8s	8'-0"	
2 - 2 x 10s	(*) 10'-0"	Rough stud opening width plus bearing on four studs
2 - 2 x 12s	(*) 12'-0"	

(*) Triple studs at jamb opening; headers to bear on 2 - 2 x 4s on each end.

Headers - Exterior Openings
Figure 7-9

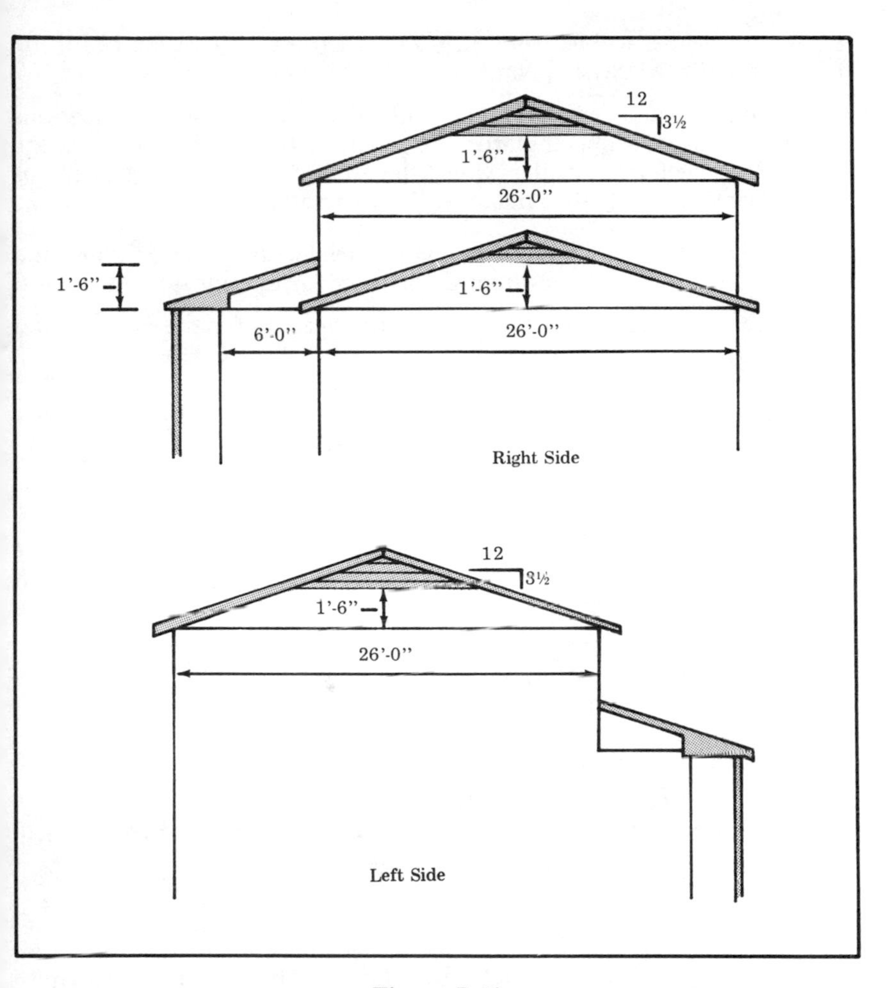

Figure 7-10

are twin or triple, an additional allowance must be made for the rough stud opening for each mullion. To illustrate: a window with a sash opening of 2'4" in width may require a rough stud opening of 2'6½" for a single unit, 5'0½" for a twin unit, and 7'6½" for a triple unit. The thickness of outside door jambs is normally 1¼" or 1$\frac{5}{16}$". Some doors are double (Figure 7-1) and some have side lights. When estimating headers for exterior openings, add 3 inches for the bearing to the rough stud opening for openings up to 8 feet and 6 inches for the bearing to the rough stud opening for openings over 8 feet wide (Figure 7-9). Example: a $\frac{2}{4}$ x $\frac{4}{2}$ twin window with a rough stud opening width of 5'1" will required 2 - 2 x 6s for the header (Figure 7-9). The bearing on two studs is 3 inches (2 x 1½") plus the rough stud opening width of 5'1", which equals 5'4". This will be the length of each header. The length of the double header will be 10'8'' (2 x 5'4''). Estimate 1 - 2 x 6 x 12 for the header for this window opening.

Manufacturers of windows and sliding doors provide specifications for the rough stud openings for their products. The specifications are available at the lumber dealers where the products are sold. Get a copy of these specifications before making your estimate.

Figure 7-11 is a form for estimating the headers for exterior walls, and Figure 7-12 is a form for estimating headers for interior door openings.

The consolidated list of material for the headers for the exterior and interior walls in Figures 7-1 and 7-2 is:

2 - 2 x 12 x 12	(48 bd. ft.)
4 - 2 x 10 x 10	(67 bd. ft.)
2 -- 2 x 8 x 8	(22 bd. ft.)
7 - 2 x 6 x 12	(84 bd. ft.)
2 - 2 x 6 x 10	(20 bd. ft.)
2 - 2 x 6 x 8	(16 bd. ft.)
9 - 2 x 4 x 12	(72 bd. ft.)
3 - 2 x 4 x 10	(20 bd. ft.)
6 - 2 x 4 x 8	(32 bd. ft.)
Total	381 bd. ft.

Note: It is recommended that this material list be given to the foreman on the job; it will prevent a lot of waste.

Temporary Wall Braces

After the exterior and interior walls are in place, temporary wall braces are needed to keep them plumb and aligned until the upper top plate and ceiling joists or trusses are installed. These temporary wall braces are normally 2 x 4 x 12s and they can be used in many places later such as blocking, nailers, drop ceilings, and lookouts. Estimating the number of 2 x 4 x 12s for the temporary wall bracing is a hit-or-miss estimate, but allowing 20 - 2 x 4 x 12s for the entire house (the braces on the first floor can be used on the second floor) will be a close estimate for most jobs.

The temporary wall braces for the first and second floors in Figures 7-1 and 7-2 will be:

Order: *20 - 2 x 4 x 12 . . . . (160 bd. ft.)*

Page_________of _________Pages					
Unit	**Rough Stud Opening**	**Add For Bearing**	**Header Length**	**Header Size (On Edge)**	**Estimate for Each Double Header**

Headers For Exterior Openings
Figure 7-11

Corner Bracing
Corner bracing adds rigidity to the structure by protecting against lateral forces such as wind acting against the walls. Either of the following may be used for corner bracing.

1. Wood sheathing installed at a 45 degree angle in opposite directions from each corner.
2. 1 x 4 inch or wider boards let into either inner or outer face of studs, sole plate and top plate located near each corner and set at approximately 45 degree angles.

Page______of______Pages					
Location	**Door Width**	**Add For Allow-ance**	**Header Length**	**Header Size (On Edge)**	**Estimate For Each Double Header**

Headers For Interior Door Openings
Figure 7-12

3. 4 x 8 plywood sheathing applied vertically.

4. $^{25}/_{32}$ inch fiberboard in 4 x 8 sheets installed vertically.

The most common practice for installing corner bracing is to use ½'' x 4' x 8' plywood sheathing at the corners and ½'' x 4' x 8' fiberboard sheathing vertically on the balance of the walls.

When estimating corner bracing, use one of the following methods:

1. If let-in-bracing is used, estimate 2 - 1 x 4 x 12s for each corner.

2. If sheet material is used, estimate 2 - 4 x 8 pieces for each corner and include with the wall sheathing estimate.

The corner bracing for the house in Figures 7-1 and 7-2 will be ½" x 4' x 8' plywood, and should be included with the wall sheathing estimate.

Wall Sheathing

Wall sheathing strengthens and adds rigidity to the exterior walls. It also adds additional insulation for the house. With energy saving on the minds of builders and home owners, special polystyrene foam boards with reflective vapor barriers that increase the "R" factor for greater energy saving are now being used by many builders. (This will be covered later under "Insulation.")

The following may be used for wall sheathing.

1. Wood board sheathing: T&G, square edge or shiplapped.
   (a) Minimum thickness: ¾ inch
   (b) Maximum width: 12 inches
   (c) Corner bracing is required unless boards are installed diagonally.

2. Plywood sheathing:
   (a) Minimum thickness: 5/16 inch (studs at 16 inch o.c.); ⅜ inch (studs at 24 inch o.c.)
   (b) Plywood sheathing is acceptable for corner bracing if installed vertically.
   (c) Nail plywood with 6d nails spaced 6 inches o.c. along edges and 12 inches o.c. along intermediate studs.

3. Fiberboard sheathing:
   (a) Minimum thickness: ½ inch
   (b) Corner bracing is required.
   (c) Nail sheathing to studs at each bearing with 1½ inch roofing nails (1¾ inch roofing nails for 25/32 inch sheathing) with the same spacing as for plywood.

4. Gypsum sheathing:
   (a) Minimum thickness: ½ inch
   (b) Corner bracing is required.
   (c) Nail sheathing to studs at each bearing with 1½ inch roofing nails with ⅜ to 7/16 inch head, 4 inches o.c. at edges and 8 inches at intermediate supports.

5. Polystyrene Foam Board Sheathing:
   (a) Follow manufacturers recommendation for installation and nail schedule.

Sheet, or panel, material has replaced wood boards for wall sheathing in most new houses. This provides for a big saving in time and labor.

When estimating wall sheathing, first check the wall section of the blueprints for the type and thickness of the material to use. Some estimators deduct for large openings such as garage doors and picture windows. The cutouts from these openings can be used as fillers around band joists and to help offset the waste for the gable sheathing. The savings in the material by deducting the openings can be lost, however, in the extra labor that will be required to cut and fit the small pieces around the windows and doors as there is less labor in sheathing the walls solid and cutting out the openings for windows and doors than in cutting and fitting small pieces of sheathing around them.

The first step in estimating wall sheathing is to calculate the total lineal feet of the perimeter of the exterior walls on each floor and multiply by the wall height for the gross area in square feet.

The second step is to calculate the gross area in square feet of the gables. Do this by multiplying the rise from the plate to the ridge (it can be scaled from the elevation on the blueprints) by the width of the gable and dividing by 2. See diagram below.

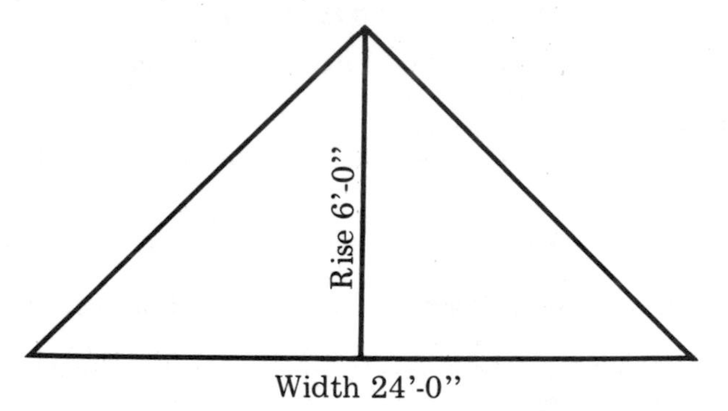

$$\text{Solution: } \frac{6 \times 24}{2} = 72 \text{ square feet}$$

Calculate the area of each gable separately (if they are different sizes) and add the totals for the total gable area for the sheathing.

The third step is to add all gross areas for the combined total gross area for the wall sheathing. If deductions for openings are made, subtract the openings from the gross area for the net area in square feet to be sheathed.

(a) If wood board sheathing is used, refer to the waste factor under subfloors in the chapter on "Floor Systems" and multiply this factor by the net area for the quantity of material to order.

(b) If 4 x 8 sheets, or panels, are used, divide the net area in square feet by 32 and round off to the next highest whole number for the number of sheets to order.

(c) If plywood corners are used for bracing, and the balance of the walls is to be fiberboard (or other types of panel), subtract the number of plywood sheets from the total number of sheets, or panels, required for the itemized list of sheathing to order.

Example: estimate the wall sheathing for the first and second floors in Figures 7-1 and 7-2 and the gables in Figure 7-10. ½" x 4' x 8' plywood will be used on all corners for bracing and ½" x 4' x 8' fiberboard sheathing will be used on the balance of the house.

1. From Figure 7-1 the perimeter of the first floor is 192'0''. From Figure 7-2 the perimeter of the second floor is 132'0''. The height of each floor is 9'0'' (8'0'' ceiling height plus 1'0'' for the floor joists).
First floor: 192' x 9' = 1728 sq. ft.
Second floor: 132' x 9' = 1188 sq. ft.

2. There are three main gables on the house (Figure 7-10). Each gable is 26'0'' wide and, because of the large vents in each gable, the rise is only 1'6''.

$$\frac{\text{Rise 1'-6'' (1.5') x width 26'-0'' (26.0'}}{2} = 19.5 \text{ sq. ft.}$$

19.5 sq. ft. x 3 gables = 58.5 sq. ft.

The 6 foot extension of the garage at the left front corner of the house has a shed-type roof with two gables. The total area for these two small shed-type gables is 9 square feet. The total gable area is:

3 main gables	58.5 sq. ft.
2 shed gables	9.0 sq. ft.
Total	67.5 or *68 sq. ft.*

Note: Gables require more waste, but the cutouts from the window and door openings will offset this waste and no additional allowance need be made.

3. The total gross area for the wall sheathing is:

First floor	1728 sq. ft.
Second floor.	1188 sq. ft.
Gables	68 sq. ft.
Total gross area	2984 sq. ft.
Deduct for garage door openings .	(126) sq. ft.
Total net area	2858 sq. ft.

$$\frac{2858 \text{ sq. ft.}}{\text{(panel size) } 32 \text{ sq. ft.}} = 89.31 \text{ or } 90 \text{ panels}$$

Total . . . . . . 90 panels
Less . . . . 20 plywood corners
70 fiberboard panels

Order:
70 - ½'' x 4' x 8' Fiberboard sheathing
20 - ½'' x 4' x 8' C-D Plywood sheathing

Nails

The nail factor per 1000 board feet of framing for studs, plates and headers is about twice that for the floor joists. There is a high percent of waste, and many nails are used for temporary use such as wall braces and scaffolds. An allowance must be made for this waste. A guideline to use in the absence of previous records is:

21 lbs. 16d common nails per 1000 bd. ft.

If records from previous jobs are kept, a more accurate nail factor can be used for each job. Example: if 129 lbs. 16d common nails were used on a previous job for the studs, plates and headers with 5980 board feet of lumber, what will be the nail factor per 1000 board feet for another job that is similar?

Answer: $129:x = 5980:1000$

$$\frac{129 \times 1000}{5980} = 21.57 \text{ or } 22$$

$x = 22$ lbs. per 1000 bd. ft.

Using the above nail factor, how many lbs. of 16d common nails will be required for the walls with 6240 board feet?

Answer: $22:x = 1000:6240$

$$\frac{22 \times 6240}{1000} = 137.28 \text{ or } 138$$

$x = 138$ lbs. 16d common nails

The total board feet of lumber for the walls in Figures 7-1, 7-2 and 7-10 is:

Sole and top plates	1432
Studs	4267
Headers	381
Temporary braces	160
Total	6240 bd. ft.

Using the nail factor of 22 lbs. 16d common nails per 1000 board feet, the quantity of nails estimated will be:

$$\frac{6240 \text{ bd. ft.} \times 22}{1000} = 137.28 \text{ or } 138$$

Order: 138 lbs. 16d common nails

There are 70 pcs. ½'' x 4' x 8' fiberboard sheathing, and the total square feet is 2240 (70' x 32' = 2240 sq. ft.). 1½'' roofing nails will be used to install the sheathing, and the nail factor is:

10 lbs. 1½" roofing nails per 1000 sq. ft.

The quantity of nails to estimate for the fiberboard sheathing is:

$$\frac{2240 \text{ sq. ft. x } 10}{1000} = 22.40 \text{ or } 23$$

Order: *23 lbs. 1½" roofing nails*

There are 20 pcs. ½" x 4' x 8' plywood sheathing for the corner bracing, and the total square feet is 640 (20' x 32' = 640 sq. ft.). 6d common nails will be used for the plywood, and the nail factor is:

11 lbs. 6d common nails per 1000 bd. ft.

The quantity of nails for the plywood will be:

$$\frac{640 \text{ sq. ft. x } 11}{1000} = 7.04 \text{ or } 7$$

Order: *7 lbs. 6d common nails*

The material list for the wall system in Figures 7-1, 7-2 and 7-10 is :

Sole and Top Plates

179 - 2 x 4 x 12	(1432 bd. ft.)

Studs

800 - 2 x 4 x 8	(4267 bd. ft.)

Headers

2 - 2 x 12 x 12	(48 bd. ft.)
4 - 2 x 10 x 10	(67 bd. ft.)
2 - 2 x 8 x 8	(22 bd. ft.)
7 - 2 x 6 x 12	(84 bd. ft.)
2 - 2 x 6 x 10	(20 bd. ft.)
2 - 2 x 6 x 8	(16 bd. ft.)
9 - 2 x 4 x 12	(72 bd. ft.)
3 - 2 x 4 x 10	(20 bd. ft.)
6 - 2 x 4 x 8	(32 bd. ft.)
Total for headers	(381 bd. ft.)

Temporary Wall Braces

20 - 2 x 4 x 12	(160 bd. ft.)

Wall Sheathing

70 - ½" x 4' x 8' fiberboard	(2240 sq. ft.)
20 - ½" x 4' x 8' C-D plywood	(640 sq. ft.)

Nails

138 lbs. 16d common nails
23 lbs. 1½'' roofing nails
7 lbs. 6d common nails

Ceiling and Roof Framing

Ceiling Joists

Ceiling joists are supports for the ceiling, and are not designed to support floor loads. If more than limited attic storage is used, floor joists are required. When roof trusses are used they replace ceiling joists. The floor plan of the blueprints will show the size, spacing and direction the ceiling joists will run. The size and spacing shown on the plans will normally be calculated by the designer of the house, but Figure 7-13 can be used as a guideline for the ceiling joist spans. The length of the joists can be determined from the floor plans by the dimensions of the rooms, and the direction the joists run. The direction the joists run is normally the shorter span.

Ceiling joists seldom span the entire width of the building as trusses do, and the direction they run varies in many houses, as shown in Figures 7-14 and 7-15. When estimating these joists, estimate each section separately.

Headers and trimmers will be required for any openings in the ceiling such as chimneys and access doors to the attic (the same as shown in Figure 6-8 for floor joists). Estimate these headers and trimmers the same as for floor joists.

Estimating ceiling joists is the same for hip or gable roofs, but the layout is different. The run of the regular joists must stop short of the outside wall to permit the hip and jack rafters to clear them. Short ceiling joists are installed perpendicular to the regular joists as fillers to permit the rafters to reach the outside wall plate (Figure 7-16).

To estimate ceiling joists, multiply the lineal feet of the wall by .75 and add one joist for the starter if the spacing is 16 inches o.c. If the spacing is 24 inches o.c., multiply the lineal feet of the wall by .50 and add one joist. This is the same as for floor joists.

If ceiling joists were used for the first and second floors for the house in Figures 7-1 and 7-2, they would be estimated as follows (all joists are on 16 inch o.c.):

First Floor (Figure 7-1)

1. Ceiling joists would be required for the ceilings in the living room, kitchen and dining room only, because the second floor is over the area from the wall separating the living room from the stairs and extends 40'0'' to the left wall of the garage.

2. The ceiling joists in the living room area will span the shortest distance, or 13'0''. Figure 7-13 shows 2 x 6 material can be used for 16 inch centers. The lineal feet of the wall is 24'2¼'' (19'0'' + 5'2¼''). The calculations are:

Nominal Size (Inches)	Spacing (Inches o.c.)	Select Structural 1950 f	Dense Construction 1700 f	Construction 1450 f	Standard 1200 f
2 x 4	12	9'- 6"	--	8'- 2"	6'- 4"
	16	8'- 6"	--	7'- 2"	5'- 6"
	24	7'- 6"	--	5'-10"	4'- 6"
2 x 6	12	14'- 4"	14'- 4"	14'- 4"	14'- 4"
	16	13'- 0"	13'- 0"	13'- 0"	12'-10"
	24	11'- 4"	11'- 4"	11'- 4"	10'- 6"
2 x 8	12	18'- 4"	18'- 4"	18'- 4"	18'- 4"
	16	17'- 0"	17'- 0"	17'- 0"	17'- 0"
	24	15'- 4"	15'- 4"	15'- 4"	14'- 4"
2 x 10	12	21'-10"	21'-10"	21'-10"	21'-10"
	16	20'- 4"	20'- 4"	20'- 4"	20'- 4"
	24	18'- 4"	18'- 4"	18'- 4"	18'- 0"

Figure 7-13
Spans
Ceiling Joists

.75 x 24.19' (24'2¼") = 18.14 (rounded off to 18)
18 + 1 = 19 ceiling joists for living room
Order: 19 - 2 x 6 x 14

3. The ceiling joists in the kitchen and dining room will span the shortest distance of 13'0". 2 x 6 material will also be used here. The lineal feet of the wall is also 24'2¼". The calculations for this area are the same as for the living room.
Order: 19 - 2 x 6 x 14

Second Floor (Figure 7-2)

4. The span of the front section of the second floor where the master bedroom and bedroom #2 are located is 15'3". Figure 7-13 shows 2 x 8 material on 16 inch centers is required here. The lineal feet of the wall is 40'0". The calculations for the ceiling joists in this section are:
.75 x 40.0' = 30
30 + 1 = 31 ceiling joists
Front section . . . . . Order: 31 - 2 x 8 x 16

5. The span of the back section of the second floor is 10'9". Figure 7-13 shows 2 x 6 material spaced on 16 inch centers may be used. The lineal feet of the wall is 40'0". The calculations for the ceiling joists in the back section are:
.75 x 40.0' = 30
30 + 1 = 31 ceiling joists
Back section . . . . . Order: 31 - 2 x 6 x 12

The ceiling joists that are required for the first and second floors in Figures 7-1 and 7-2 are given below Figure 7-17 on page 30.

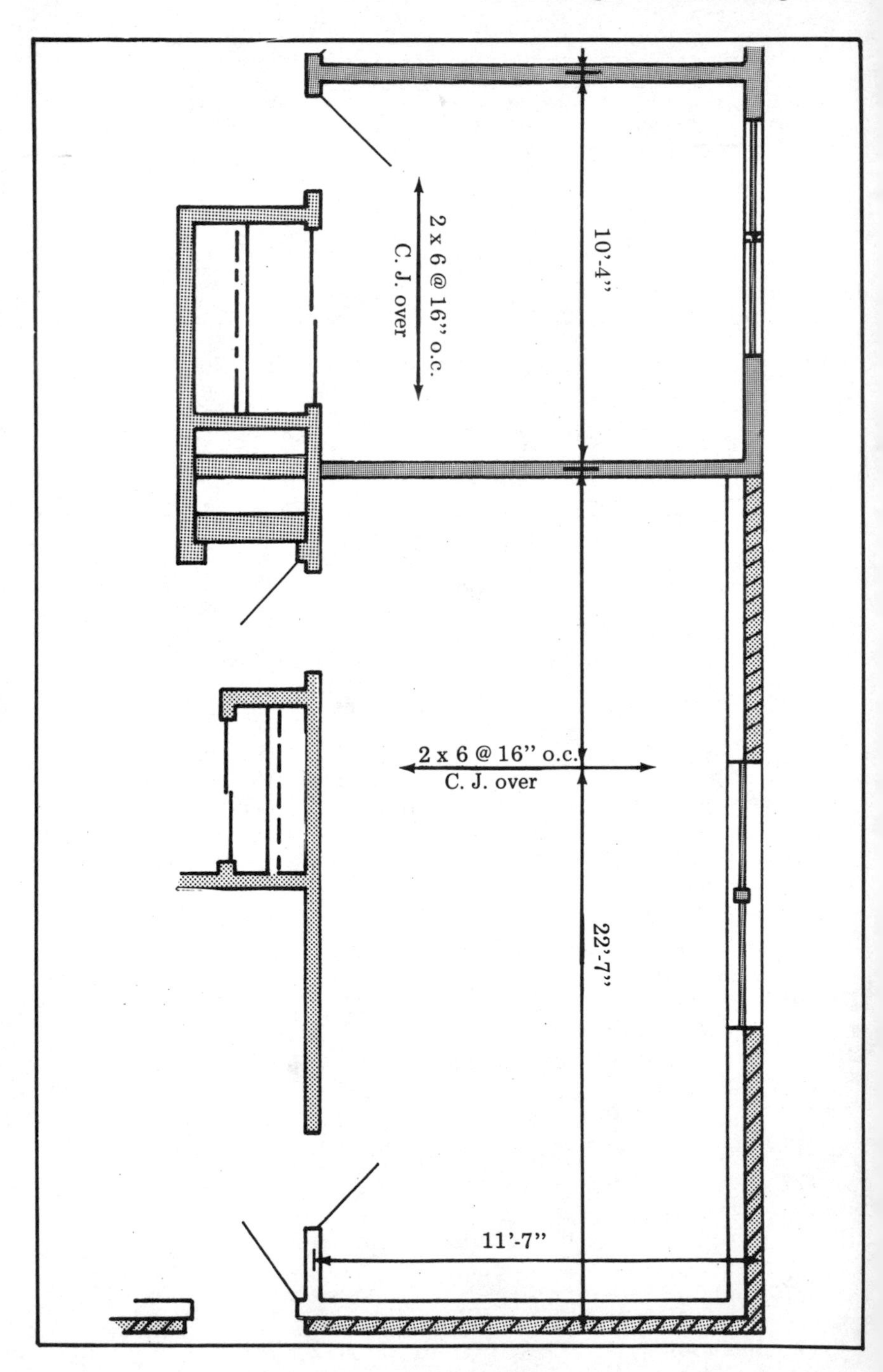

Figure 7-14
Ceiling Joists Running Two Directions

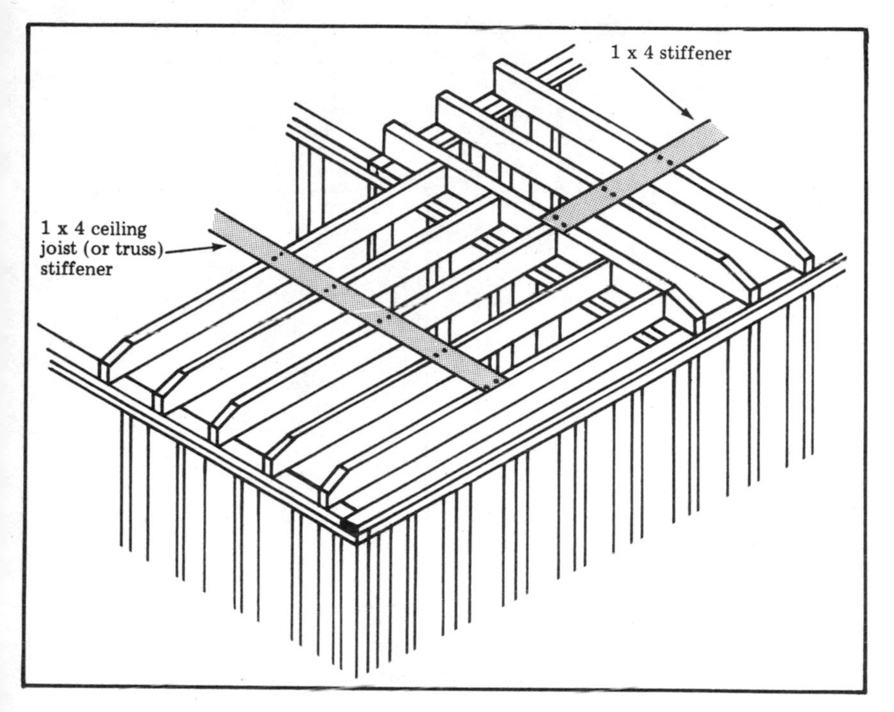

Figure 7-15
Ceiling Joists Running Two Directions

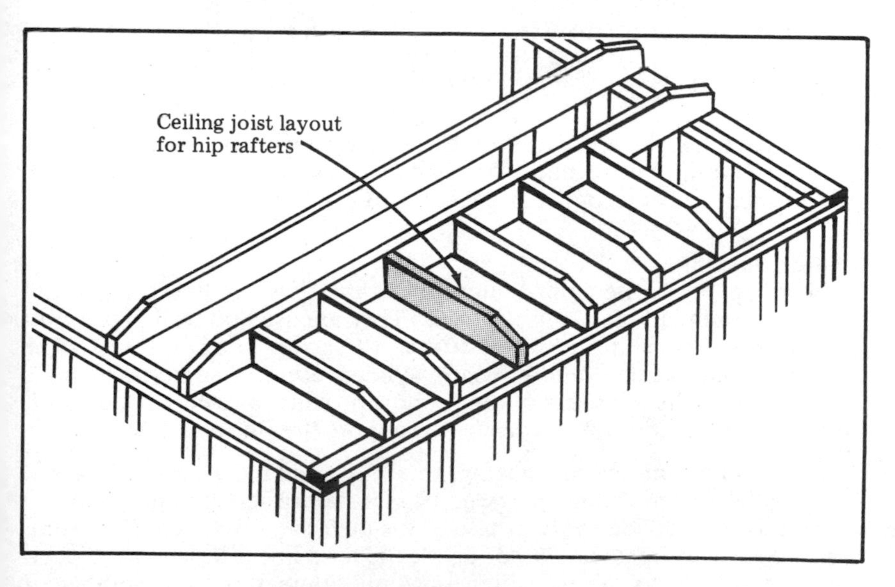

Ceiling Joists for Hip Roof
Figure 7-16

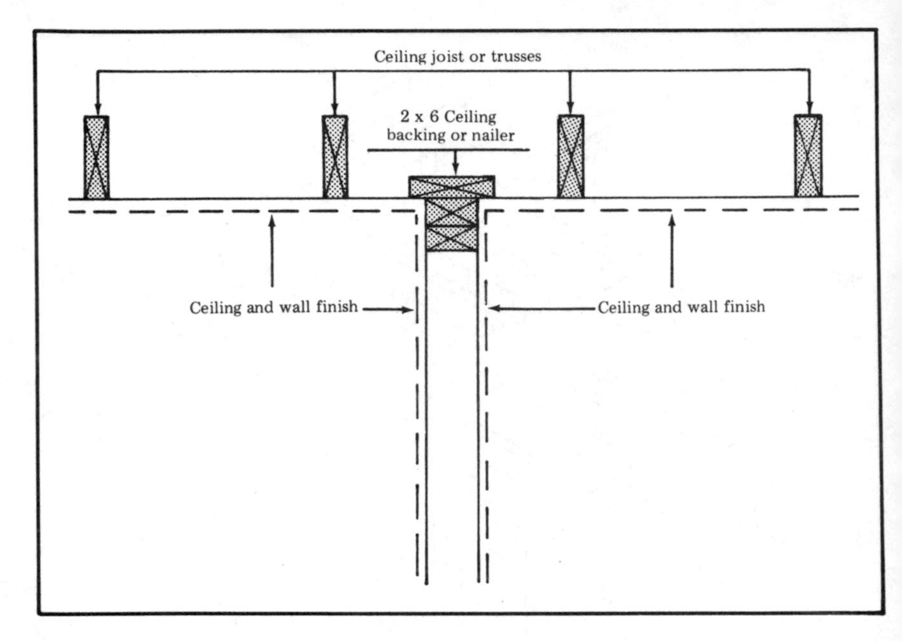

Figure 7-17
Wall Parallel to Ceiling Joists

First floor: living room	19 - 2 x 6 x 14 (266 bd. ft.)
Kitchen dining room	19 - 2 x 6 x 14 (266 bd. ft.)
Second floor: front	31 - 2 x 8 x 16 (661 bd. ft.)
back	31 - 2 x 6 x 12 (372 bd. ft.)
	Total (1565 bd. ft.)

Note: Roof trusses are used for the house shown in Figure 7-2. The above calculations were made for when rafters and ceiling joists are used. The estimate for the roof trusses is shown later in this chapter.

Ceiling Backing

Ceiling backing is the nailing support for the finish ceiling for gypsum board, gypsum lath and ceiling panels. They are used on wall plates running parallel to ceiling joists or trusses as shown in Figure 7-17. The backing should be continuous. Many builders use 2 x 6 material on a 2 x 4 wall because the thickness of the framing provides a firm base on which to nail, and the width is sufficient to support the finish ceiling.

When estimating ceiling backing, total the lineal feet of all walls running parallel to the ceiling joists or trusses from the floor plans of the blueprints, and divide by the length of the material to be used. For example, if 2 x 6 x 12s are being used, divide the total lineal feet by 12. Round this number off to the next highest whole number for the number of pieces required.

The total lineal feet of the walls running parallel to the trusses for the first and second floors in Figures 7-1 and 7-2 is:

First Floor	87
Second Floor	105
Total	192 lin. ft.

The material used for the ceiling backing will be 2 x 6 x 12s. The quantity of material is calculated as follows:

$$\frac{192 \text{ (lin. ft.)}}{12} = 16 \text{ pieces}$$

Ceiling Backing Order: *16 - 2 x 6 x 12 (192 bd. ft.)*

Note: Some waste material may be used for the ceiling backing, and if some of the ceiling joists fall over a parallel wall, they can be moved enough to provide a nailing base for the ceiling finish.

Ceiling Joist Stiffener
These nailers are used to hold and stiffen ceiling joists or trusses and to hold them in alignment (Figure 7-15). 1 x 4 material is sufficient. It is nailed to each ceiling joist or truss with two 7d or 8d nails at the approximate center of the span in each section. Estimate by adding the lineal feet of each section. Example: the total lineal feet of 1 x 4s required for the first and second floors in Figures 7-1 and 7-2 is:

First Floor (Figure 7-1)

Front section	24'-2¼"
Back section	24'-2¼"

Second Floor (Figure 7-2)

Front section	40'-0 "
Back section	40'-0 "
	128'-4½" (*)

(*) Allowing for waste, round off to 140'0"
Ceiling Joist Stiffeners . . . . . Order: 140 lin. ft. 1 x 4

Rafters
Rafters are the supporting members of the roof system. The size and spacing of the rafters must be in accordance with recognized engineering analysis procedures. There are different types of rafters, and the most common types are shown and explained in Figure 7-18.

(*) The combined jack and common rafter should be the same length as the common rafters; the hip and valley rafters will then be at a 45 degree angle with the common rafters.

Common Rafters: Run from the ridge to the plate.
Hip Rafters: Run from the ridge to the plate at a 45 degree angle to the common rafters at the outside corners.
Valley Rafters: Run from the ridge to the plate at a 45 degree angle to the common rafters at the inside corners.

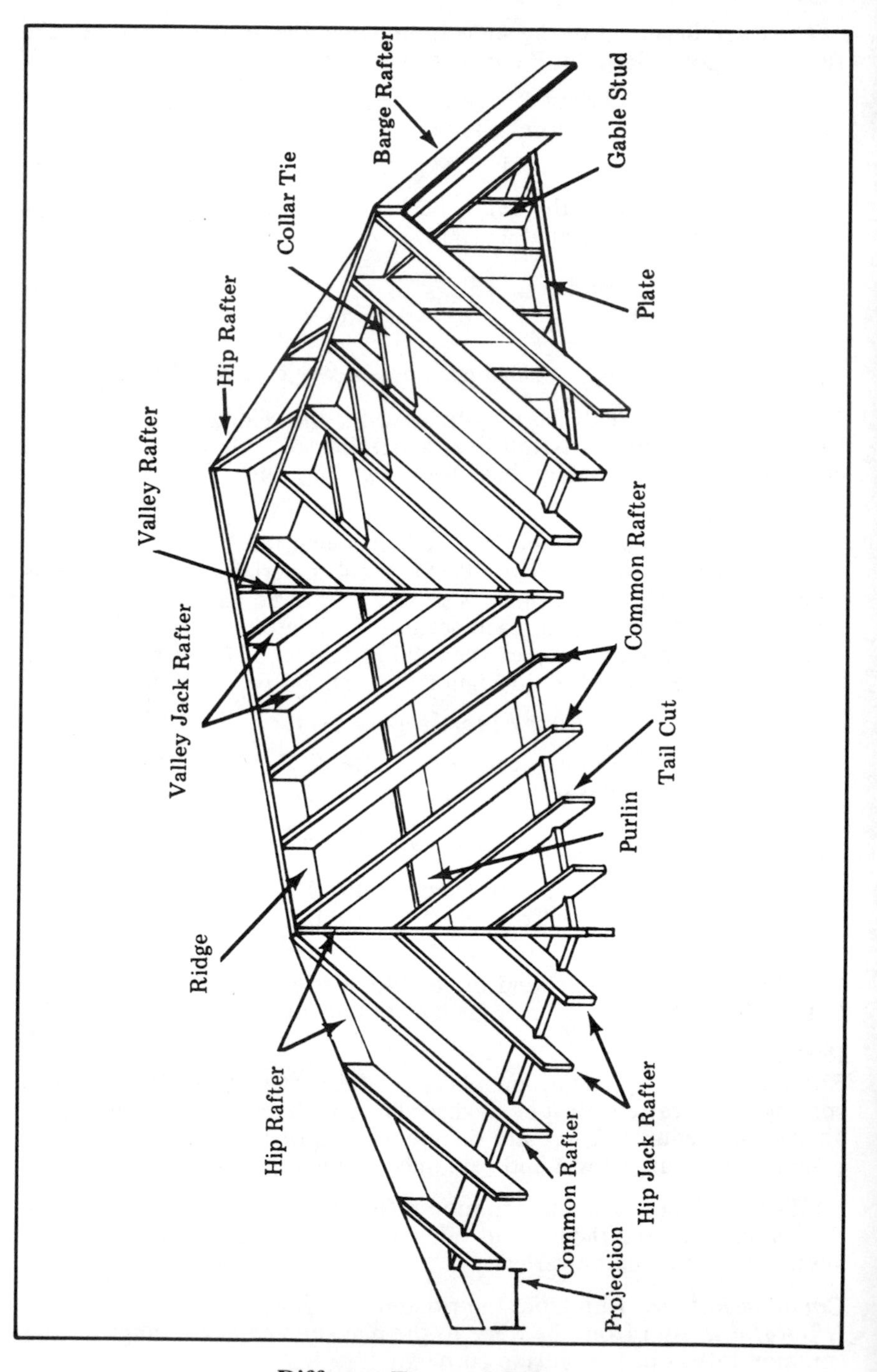

Different Types of Rafters
Figure 7-18

Nominal Size (inches)	Spacing (inches o.c.)	Select Structural	Dense Construction	Construction	Standard
		1950 f	1700 f	1450 f	1200 f
2 x 4	12	11'6"	----------	9'6"	7'4"
	16	10'6"	---------	8'4"	6'4"
	24	9'2"	---------	6'10"	5'2"
2 x 6	12	16'10"	16'10"	16'10"	16'10"
	16	15'8"	15'8"	15'8"	15'0"
	24	13'10"	13'10"	13'6"	12'2"
2 x 8	12	21'2"	21'2"	21'2"	21'2"
	16	19'10"	19'10"	19'10"	19'10"
	24	17'10"	17'10"	17'10"	16'8"

Figure 7-19
Rafter Spans

Jack Rafters: Run from the hip or valley rafters to the plate or ridge.
Cripple Rafters: Run between the hip and valley rafters.
Ridge: The highest horizontal member in the roof system and is an aid in the installation of the rafters.

Common Rafters
Common rafters run from the ridge to the plate as shown in Figure 7-18. The span of the rafters in Figure 7-19 is maximum clear span, or the actual length of members between inner faces of support, measured along the slope. The size and spacing of the common rafters are normally shown on the wall section of the blueprints.

Figure 7-20 shows the terminology used in calculating common rafter lengths.

Span: the width of framing members of the building. *Run:* one-half the span. For computing the rafter overhang, it is the distance from the wall to the tail cut of the rafter.*Rise:* the vertical distance from the plate line to the measuring line (sloping dotted line) vertex. *True Rafter Length:* the distance between the face of the ridge and the outer face of the wall framing. *Net Rafter Length:* the distance between the face of the ridge and the tail cut, or the total length of the rafter. *Factor:* the length of the common rafter in inches per foot run.

The rise of the roof is expressed as a number of inches per one foot run and is normally designated on the elevation section of the blueprints as a symbol. Example:

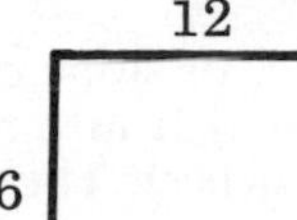

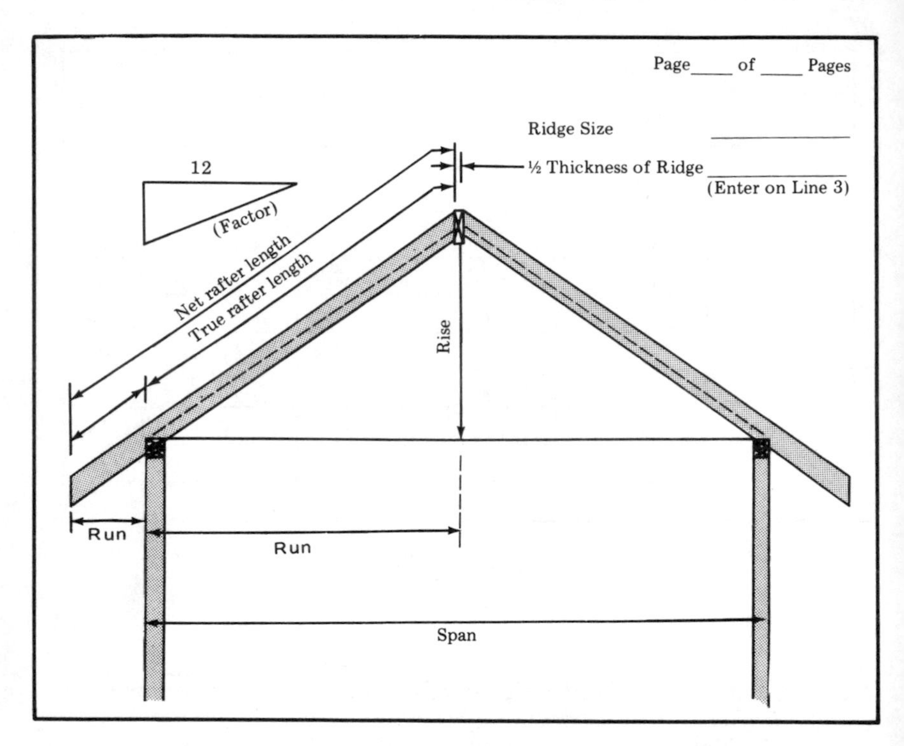

Figure 7-20
Common Rafters

1. ______ x ______ = ______________________________ inches
   Run — Factor — Rafter length from plate to center or ridge (Enter on line 2)

2. ___ ______ inches divided by 12 = _______ feet (_____ feet _____ inches)
   (From line 1) — Enter on line 3 (See conversion chart)

3. ________ feet ______ inches less ____________ inches = _____ feet _____ inches
   (From line 2) — (½ thickness of ridge) True rafter length (Enter on line 6)

4. Rafter overhang: _________ x _________ = _________ inches
   run — Factor — (Enter on line 5)

5. _________ inches divided by 12 = ______ feet (________ feet ______ inches)
   (From line 4) — (See conversion chart)

6. _________ feet _____ inches plus overhang _____ feet _____ inches = _____ feet _____ inches
   True rafter length (From line 3) — (From line 5) — Net rafter length

Figure 7-20A
(Worksheet)

This is referred to as the pitch, or slope of the roof. (Pitch and slope are synonymous to most builders.) It means the roof rises 6 inches for each foot run (expressed as 12 inches). The rafter length per foot run is

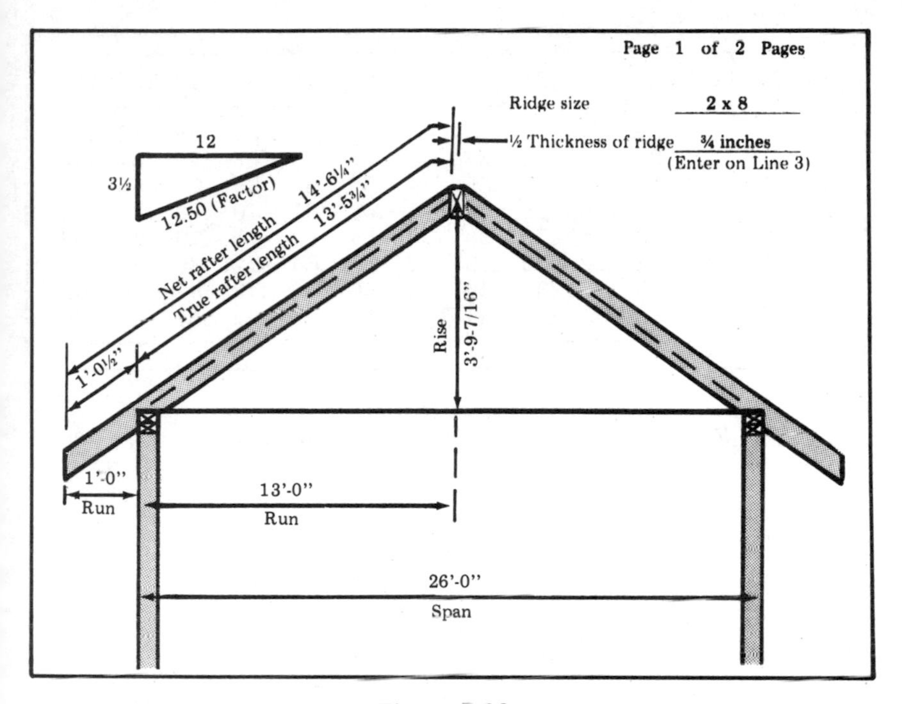

Figure 7-20
Common Rafters
Sample

1. 13' (Run) x 12.50 (Factor) = 162.5 inches (Rafter length from plate to center or ridge (Enter on line 2))
2. 162.5 (From line 1) inches divided by 12 = 13.54 feet (13 feet 6½ inches) (Enter on line 3 (See conversion chart))
3. 13 (From line 2) feet 6½ inches less 3/4 (½ thickness of ridge) inches = 13 feet 5¾ inches (True rafter length (Enter on line 6))
4. Rafter overhang: 1'-0" (run) x 12.50 (Factor) = 12.5 inches (Enter on line 5)
5. 12.5 (From line 4) inches divided by 12 = 1.04 feet (1 feet 0½ inches) (See conversion chart)
6. 13 feet 5¾ inches (True rafter length (From line 3)) plus overhang 1 feet 0½ inches (From line 5) = 14 feet 6¼ inches (Net rafter length)

Figure 7-20A
(Worksheet)
Sample

4th	8th	0"	1"	2"	3"	4"	5"	6"	7"	8"	9"	10"	11"
0	0	.00	.08	.17	.25	.33	.42	.50	.58	.67	.75	.83	.92
	1	.01	.09	.18	.26	.34	.43	.51	.59	.68	.76	.84	.93
1	2	.02	.10	.19	.27	.35	.44	.52	.60	.69	.77	.85	.94
	3	.03	.11	.20	.28	.36	.45	.53	.61	.70	.78	.86	.95
2	4	.04	.13	.21	.29	.38	.46	.54	.63	.71	.79	.88	.96
	5	.05	.14	.22	.30	.39	.47	.55	.64	.72	.80	.89	.97
3	6	.06	.15	.23	.31	.40	.48	.56	.65	.73	.81	.90	.98
	7	.07	.16	.24	.32	.41	.49	.57	.66	.74	.82	.91	.99

Figure 7-21
Decimal Equivalents in Hundredths of Fractional Parts of a Foot

calculated as follows:

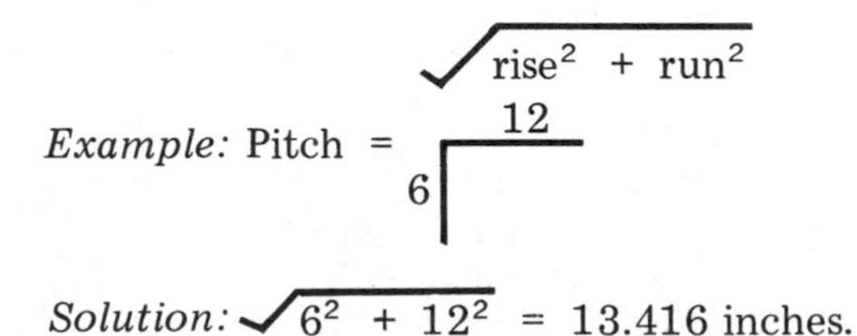

$$\sqrt{\text{rise}^2 + \text{run}^2}$$

Example: Pitch = 6 / 12

Solution: $\sqrt{6^2 + 12^2} = 13.416$ inches.

Figures 7-20 and 7-20A are forms that may be used for calculating rafter lengths using factors for different roof pitches. Sample worksheets show how these computations are made. These worksheets are precise for computing the rafter lengths for cutting the top or plumb cut, the bird's mouth and the tail cut of rafters. Figure 7-21 is a table of decimal equivalents in hundredths of fractional parts of a foot. Figure 7-23 is a table of decimal equivalents of fractional parts of an inch. Each of these tables supplements Figures 7-20 and 7-20A.

Figure 7-22 is a table showing the lengths (or factors) per foot run for common rafters, jack rafters, and hip and valley rafters.

Some estimators scale the rafter lengths from the blueprints, but blueprints shrink, so this method is not accurate. Example: the estimator may scale 13'10" for the rafter length when the net rafter length is 14'4". Thus, he would order the rafters in lengths of 14'0" when they should be ordered in lengths of 16'0". Figure 7-24 is a table of ratios the estimator may use as a fast and reasonably accurate method for estimating rafter lengths for the purpose of ordering the material. The ratios in Figure 7-24 are used as follows:

1. Use the ratio for the roof pitch of the house.

2. Multiply the ratio by the run of the house plus the overhang (Figure 7-20).

Pitch of Roof	Common Rafters: Length for each foot of runs in inches	Jack Rafters: Length of shortest jack on 16 inch centers	Jack Rafters: Length of shortest jack on 24 inch centers	Hip and Valley Rafters: Length for each foot of run of common rafters in inches
12				
1/12	12.042			
1½/12	12.093			
2/12	12.166			
2½/12	12.258			
3/12	12.369	16.485	24.740	17.234
3½/12	12.500	16.659	25.001	17.328
4/12	12.649	16.858	25.299	17.436
5/12	13.000	17.326	26.001	17.692
6/12	13.416	17.880	26.833	18.000
7/12	13.892	18.516	27.786	18.358
8/12	14.422	19.222	28.845	18.762
9/12	15.000	19.991	29.999	19.209
10/12	15.620	20.820	31.242	19.698
11/12	16.279	21.697	32.558	20.224
12/12	16.971	22.620	33.942	20.785

Figure 7-22
Factors Per Foot Run

Example: From the sample worksheet for Figure 7-20, the roof pitch is 3½/12.

1. The ratio for a 3½/12 roof pitch is 1.042.

2. The run of the house plus the roof overhang is 14'0'' (*) (13'0'' + 1'0'')

(*) All calculations should be in feet and decimal equivalents rather than in feet and inches.

14.0' x 1.042 = 14.59'
14.59' is 14'7⅛'' (Refer to conversion chart in Figure 7-21.)

The rafter length from the above calculation is 14'7⅛'' and the precise net rafter length calculated in the sample worksheet for Figure 7-20 is 14'6¼''. The rafters will have to be ordered in lengths of 16'0'' for either calculation. For speed, the estimator may use the ratios in Figure 7-24 for estimating rafter lengths for most residential houses.

When estimating common rafters take the lineal feet of the eave (this includes the overhang) perpendicular to the rafters and multiply by .75 and add one for 16 inch o.c. spacing, or by .50 and add one for 24 inch o.c. spacing.

Example: If rafters were used in the roof system for the first and second floors in Figure 7-1, 7-2 and 7-10 and they were spaced 16 inch o.c.,

Fourths	Eighths	Sixteenths	Thirty-Seconds	Decimal Equivalents
			1/32	.031
		1/16	2/32	.063
			3/32	.094
	1/8	2/16	4/32	.125
			5/32	.156
		3/16	6/32	.188
			7/32	.219
1/4	2/8	4/16	8/32	.250
			9/32	.281
		5/16	10/32	.313
			11/32	.344
	3/8	6/16	12/32	.375
			13/32	.406
		7/16	14/32	.438
			15/32	.469
2/4	4/8	8/16	16/32	.500
			17/32	.531
		9/16	18/32	.563
			19/32	.594
	5/8	10/16	20/32	.625
			21/32	.656
		11/16	22/32	.688
			23/32	.719
3/4	6/8	12/16	24/32	.750
			25/32	.781
		13/16	26/32	.813
			27/32	.844
	7/8	14/16	28/32	.875
			29/32	.906
		15/16	30/32	.938
			31/32	.969

Figure 7-23
Decimal Equivalents of Fractional Parts of an Inch

the number of rafters required would be calculated as follows:

1. The roof overhang is 1'-0'' for each gable end. From Figure 7-2 (second floor) the length of the wall perpendicular to the rafters is 40'-0''. The lineal feet of the eave will be 40'-0'' plus 1'-0'' overhang on each gable end; a total of 42'-0''.

42.0' x .75 = 31.50 or 32
32 + 1 = 33 rafters for each section
33 x 2 = 66 rafters for second floor

Roof Pitch	Factor	Divided By	Ratio
1/12	12.042	12	1.004
1½/12	12.093	12	1.008
2/12	12.166	12	1.014
2½/12	12.258	12	1.022
3/12	12.369	12	1.031
3½/12	12.500	12	1.042
4/12	12.649	12	1.054
5/12	13.000	12	1.083
6/12	13.416	12	1.118
7/12	13.892	12	1.158
8/12	14.422	12	1.202
9/12	15.000	12	1.250
10/12	15.620	12	1.302
11/12	16.279	12	1.357
12/12	16.971	12	1.414

Figure 7-24
Ratio of Common Rafter Length to Factors Per Foot Run

2. The length of the wall for the kitchen, dining and living rooms on the first floor in Figure 7-1 is 24'0''. This is the building length (64'0'') less the length of the second floor (40'0''). The lineal feet of the eave will be 24'-0'' + 1'-0'' overhang, or 25'-0''.

25.0' x .75 = 18.75 or 19
19 + 1 = 20 rafters for each section
20 x 2 = 40 rafters for first floor.

Note: The rafter that would rest on the gable can be omitted and the material used for the overhang lookouts if the lookouts are spaced 24 inches o.c. If the length of each lookout does not exceed 2'0'', no extra material will have to be ordered for the overhang on the gable ends.

3. The sample worksheet for Figure 7-20 (7-20A) shows the net rafter length to be 14'6¼''. A check of the span tables in Figure 7-19 shows that 2 x 6 on 16 inch centers may be used. The material that will be ordered will be in lengths of 16'0''. The rafters that will have to be ordered will be:

First Floor	40	- 2 x 6 x 16
Second Floor	66	- 2 x 6 x 16
Total	106	- 2 x 6 x 16 (1696 bd. ft.)

Hip and Valley Rafters

Hip rafters run from the ridge to the plate at a 45 degree angle to the common rafters at the outside corners. Valley rafters run from the ridge to the plate at a 45 degree angle to the common rafters at the inside corners (Figure 7-18).

As the roof pitch of common rafters is designated as the inches of rise per foot run (example: 6 | 12), hip and valley rafters are designated as 6 | 17

The number 17 is to the hip and valley rafter as 12 is to the common rafter. The number 17 is calculated as follows:

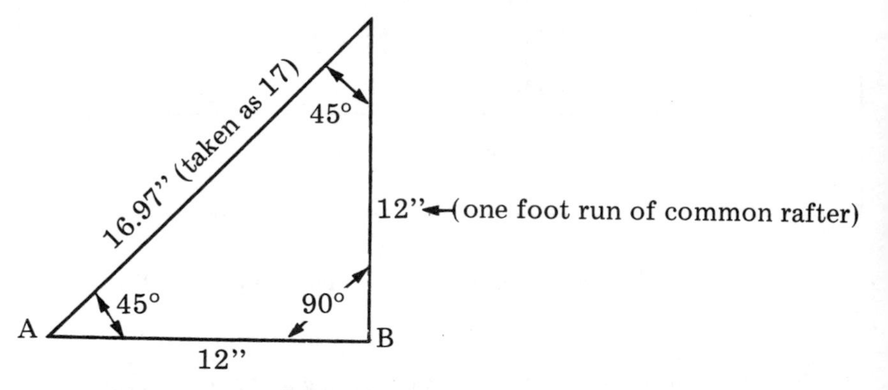

In a right triangle the sides opposite 45 degree angles are equal. Because the length of the hip and valley rafter is calculated per foot run of the common rafter, the solution is:

$$\sqrt{12^2 + 12^2} = 16.97 \ (*)$$

(*) This is shown in section "H", Figure 7-25.

If the roof pitch was 8 | 12 the length of the hip rafter per foot run of the common rafter in Figure 7-25 would be:

8 | 16.97

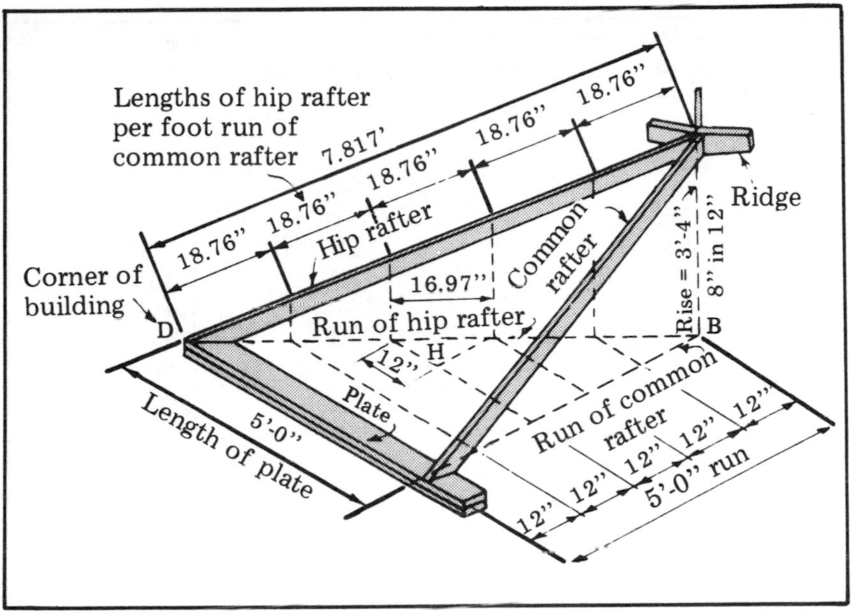

Figure 7-25

$$\sqrt{8^2 + 16.97^2} = 18.76 \text{ (Factor)}$$

If the run of the common rafter was 5'-0'', the length of the hip rafter as shown in Figure 7-25 would be:

Length of hip or valley rafter = run of common rafter x factor

Solution: 5.0' x 18.76'' = 93.80''

$$\frac{93.80''}{12} = 7.817' \ (*)$$

(*) Add the rafter overhang to this length.

Figure 7-22 shows the length of the hip and valley rafters in inches per foot run of the common rafter for different roof pitches.

To estimate the length of hip and valley rafters, refer to Figure 7-26, which shows the ratio of hip and valley rafter lengths to the run of common rafters for different roof pitches. To compute the length of hip and valley rafters from the ratios in Figure 7-26, do the following:

1. Select the ratio for the roof pitch of the house.
2. Multiply the ratio by the run of the common rafter. Example: in Figure 7-25 the roof pitch is 8/12 and the run of the common rafter is 5'0''. From Figure 7-26, we see that the ratio for a 8/12 roof pitch is 1.564.

1.564 x 5.0' = 7.820'(*)

(*) This length is to the plate only. If there is an overhang, add the run of

Roof Pitch	Factor	Divided By	Ratio
1/12	--	--	--
1½/12	--	--	--
2/12	--	--	--
2½/12	--	--	--
3/12	17.234	12	1.436
3½/12	17.328	12	1.444
4/12	17.436	12	1.453
5/12	17.692	12	1.474
6/12	18.000	12	1.500
7/12	18.358	12	1.530
8/12	18.762	12	1.564
9/12	19.209	12	1.601
10/12	19.698	12	1.642
11/12	20.224	12	1.685
12/12	20.785	12	1.732

Figure 7-26
Ratio of Hip and Valley Rafter Length to Run of Common Rafter

the overhang to the run of the common rafter for the total run, then multiply the total run by the ratio for the total length of the hip or valley rafter. Example: if there was an overhang of 1'0'' for the rafters in Figure 7-25, the total run would then be 6'0'' (5'0'' + 1'0''). The total hip rafter length would be:

6.0' x 1.564 = 9.384' (9'-4⅝'')

When estimating the number of hip and valley rafters, check the elevation section of the blueprints for the number shown on the roof plan.

Jack Rafters

Any rafter that does not extend from the plate to the ridge is called a jack rafter (Figure 7-18). They may be classified as:

1. Hip Jacks: running from the hip rafter to the plate.
2. Valley Jacks: running from the valley rafter to the ridge.
3. Cripple Jacks: running from the hip rafter to the valley.

Figure 7-22 shows the length of the shortest jack rafter on 16 inch o.c., and the length of the shortest jack rafter on 24 inch o.c. These calculations are made as follows:

Rafters Spaced 16 inch o.c.

12

Roof Pitch 8

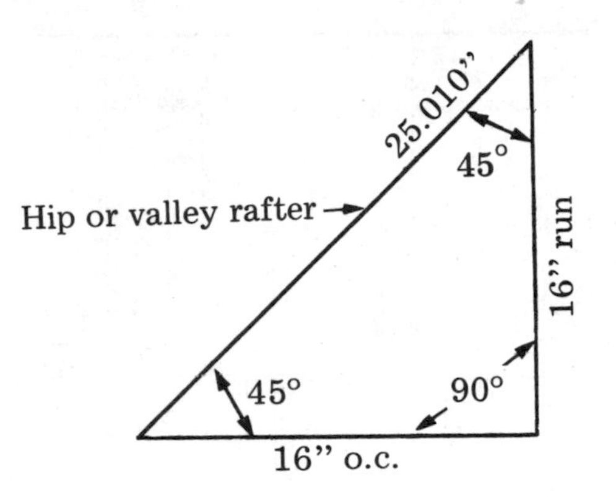

The factor for hip and valley rafters for a roof pitch of 8/12 from Figure 7-22 is 18.762.

Length of Hip and Valley Rafter = Hip and Valley Factor x Run (in feet)

18.762 x 1.333' (16") = 25.010"

Length of hip rafter per 16" run of common rafter

25.010"

Length of Jack Rafter will be:

$$\sqrt{25.010^2 - 16^2} = 19.222''$$

The length of the shortest jack rafter spaced 16 inches o.c. for a roof pitch of 8/12 is:

19.222"

If the rafters are spaced 24 inch o.c. the calculations will be:

Hip and Valley Rafter Factor 18.762
Length of hip rafter: 18.762 x 2.0' (24" run)
18.762 x 2.0' = 37.524'

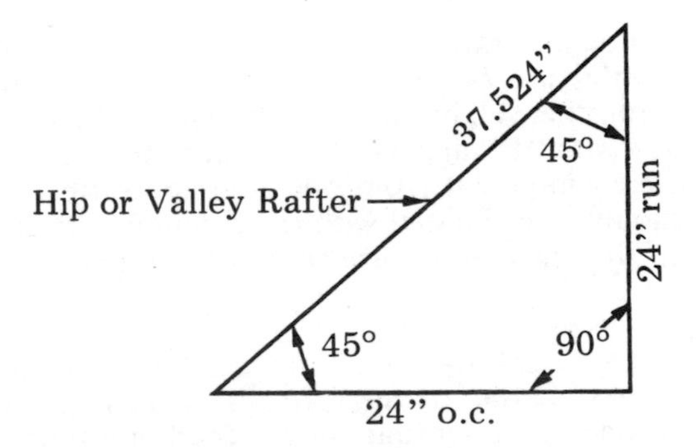

Length of Jack Rafter will be:

$$\sqrt{37.524^2 - 24^2} = 28.845''$$

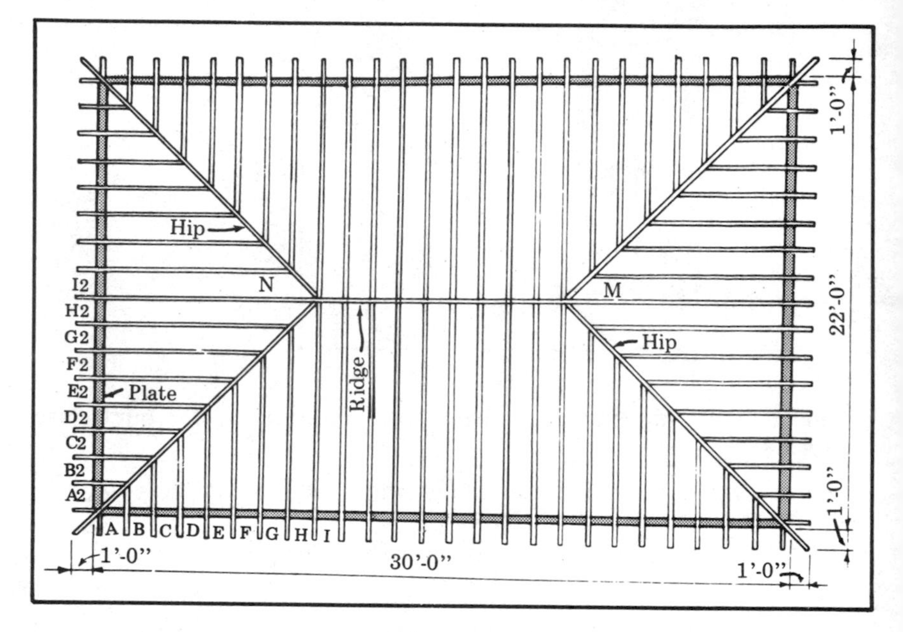

Figure 7-27
Framing Detail for a Hip Roof

The length of the shortest jack rafter spaced 24 inches o.c. for a roof pitch of $\frac{8}{12}$ is:

28.845"

The length of the second jack rafter will be twice that of the shortest jack, the third jack rafter will be three times the length and so on. This is true for both 16 inch o.c. and 24 inch o.c. spacing.

When estimating jack rafters it is not necessary to calculate each length separately. From Figure 7-27 jack rafters "A" or "A_2" and jack rafters "H" or "H_2" together can be cut from the length of one common rafter. The same is true with "B" and "G", "C" and "F", and "D" and "E". Two jack rafters can be cut from one common rafter. The number of jack rafters should be estimated with the common rafters and two extra rafters per hip for waste allowance should be added.

Ridge

The ridge board is the highest horizontal member in the roof system, and is an aid in the installation of the rafters (Figure 7-18). The size of the ridge should be one size wider than the rafters. For example, if the rafters are 2 x 6, the ridge should be 2 x 8.

The length of the ridge on a straight gable roof is the length of the building plus the overhangs. On intersecting gable roofs the ridge length is the length of the intersection plus the extension of the ridge to the adjacent roof (Figure 7-32). Normally this ridge extension is equal to the total length of the common rafter on the building extension. For example, if the building extension, or offset, is 16'0'' and the length of the common rafter on the extension is 10'0'', the length of the ridge for the extension should be 26'0''.

The length of the ridge on a full hip roof is the length of the building, less its width (Figure 7-31). On intersecting hip roofs the length of the ridge will be the length of the intersection plus the overhang.

Many estimators scale the ridge length. If the plans are drawn to a true scale, this method may be accurate enough to order the material. However, an allowance should be made for any shrinkage of the blueprints.

The length of the ridge for the house in Figures 7-1, 7-2 and 7-10 is:

First floor		25'-0"
Second floor		42'-0"
	Total	67'-0"

Order: *5 - 2 x 8 x 14 (93 bd. ft.)*

Collar Beam

Collar beams are horizontal framing members which tie rafters together to prevent roof thrust (Figure 7-28). They are located in the upper third of the attic space below the ridge. The maximum spacing of collar beams is 4 feet o.c.

If collar beams are spaced 4 feet o.c., divide the length of the house by 4 to determine the number required. (Do not include the overhang.) Example: the number of collar beams required for the house in Figures 7-1, 7-2 and 7-10 will be:

First floor length	24'-0"
Second floor length	40'-0"
Total	64'-0"

$$\frac{64}{4} = 16 \text{ collar beams}$$

Normally material 8'0'' in length will make one collar beam.
Order: 16 -2 x 4 x 8 (85 bd. ft.)

Rafter Supports

Rafter supports, or rafter braces, are used to prevent the roof from sagging under the weight of the dead and live loads the rafters must support. For maximum support, they should be 2 x 4s and have the same spacing as the rafters. They should always run from the rafter to a bearing partition (Figure 7-28). The length of the rafter supports varies with the different roof pitches.

In Figure 7-10 the length of each rafter support will be 6'0''. The number of rafters, not including the roof overhang, for the house is 100.

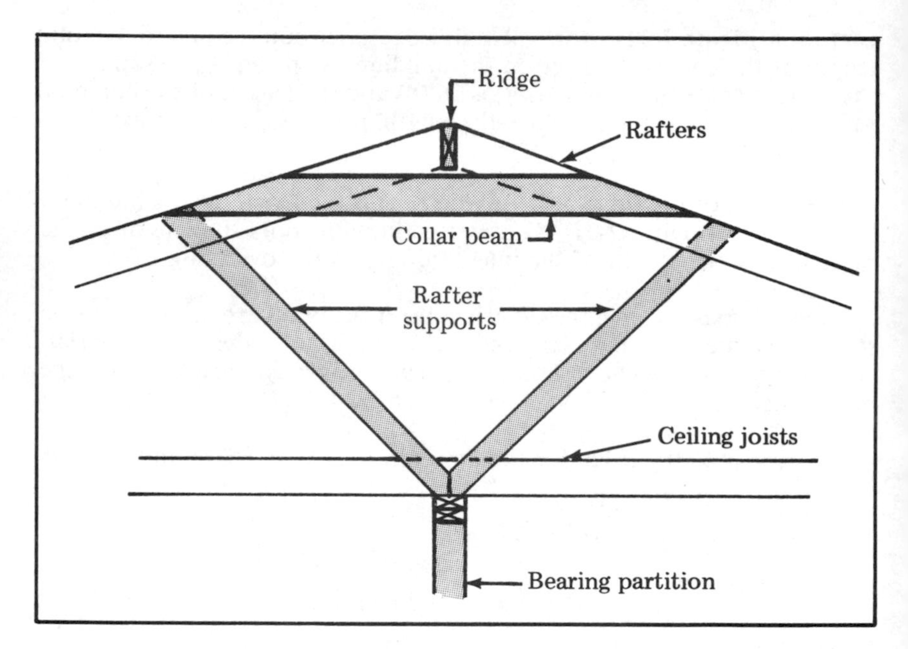

Figure 7-28
Collar Beam and Rafter Supports

The number of roof supports will then be 100 - 2 x 4s. If each roof support is 6'0'' in length, the amount of material required will be:

$$\frac{100 \times 6}{12} = 50 \text{ pieces}$$

Order: *50 — 2 x 4 x 12 (400 bd. ft.)*

Lookouts

Lookouts are short framing members nailed to the sides of rafters and extending to the wall of the house. They provide the soffit for the roof overhang (Figure 7-29). There should be one lookout per rafter, including the gable overhang. The length of the lookouts is shown on the wall section of the blueprints. The length of the eave will be the lineal feet of the nailer. The material normally used for lookouts is 2 x 4.

The material for the lookouts for Figures 7-1 and 7-2 is calculated as follows:

1. The number of rafters for the first and second floors is 106. Therefore, 106 lookouts will be required.
2. From the wall section of the blueprints the overhang is determined to be 1'0''.
3. The length of the eaves for the front and back is 134 lineal feet.

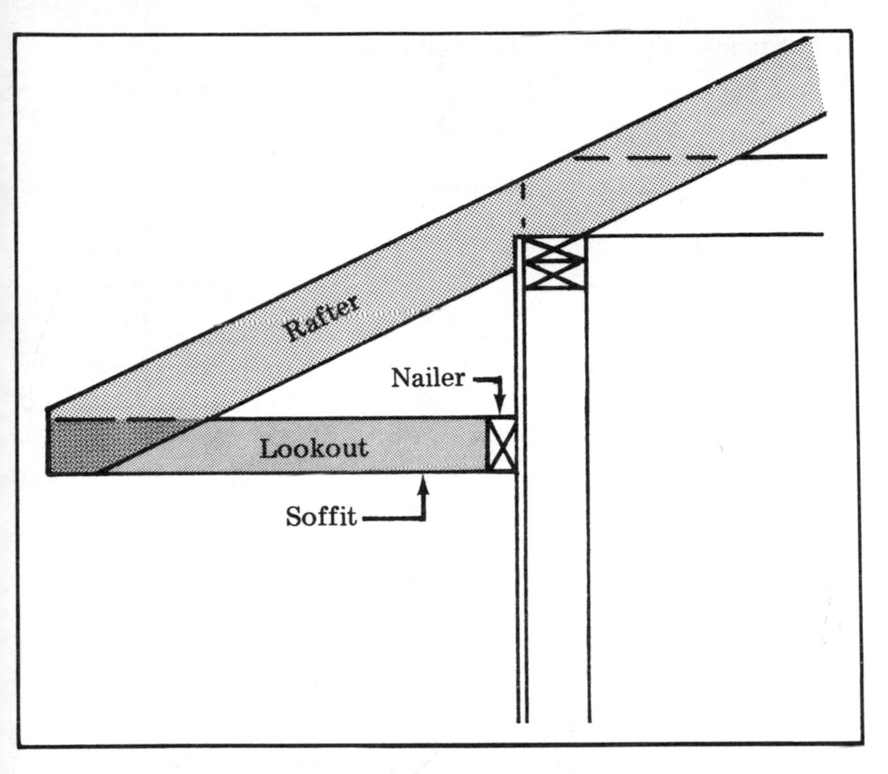

Figure 7-29
Framing for Roof Overhang Soffit

Lineal feet of lookouts	106 x 1.0' = 106
Lineal feet of nailers	134
Total	240 lin. ft.

$$\frac{240}{12} = 20 \text{ pieces}$$

Order: *20 — 2 x 4 x 12 (160 bd. ft.)*

Roof Sheathing
Plywood has almost completely replaced wood boards for roof sheathing in residential construction. One-half inch plywood is acceptable for rafter spacing of 16 inch o.c. or 24 inch o.c. In regions where heavy snow loads are common use ⅝ inch plywood. Use 6d threaded or 8d common nails and nail the plywood at each bearing 6 inches o.c. along all edges and 12 inches o.c. along intermediate members.

Before ordering roof sheathing, compute the roof area for the quantity of sheathing needed. Most roof sections in residential construction have one or more of the following shapes.

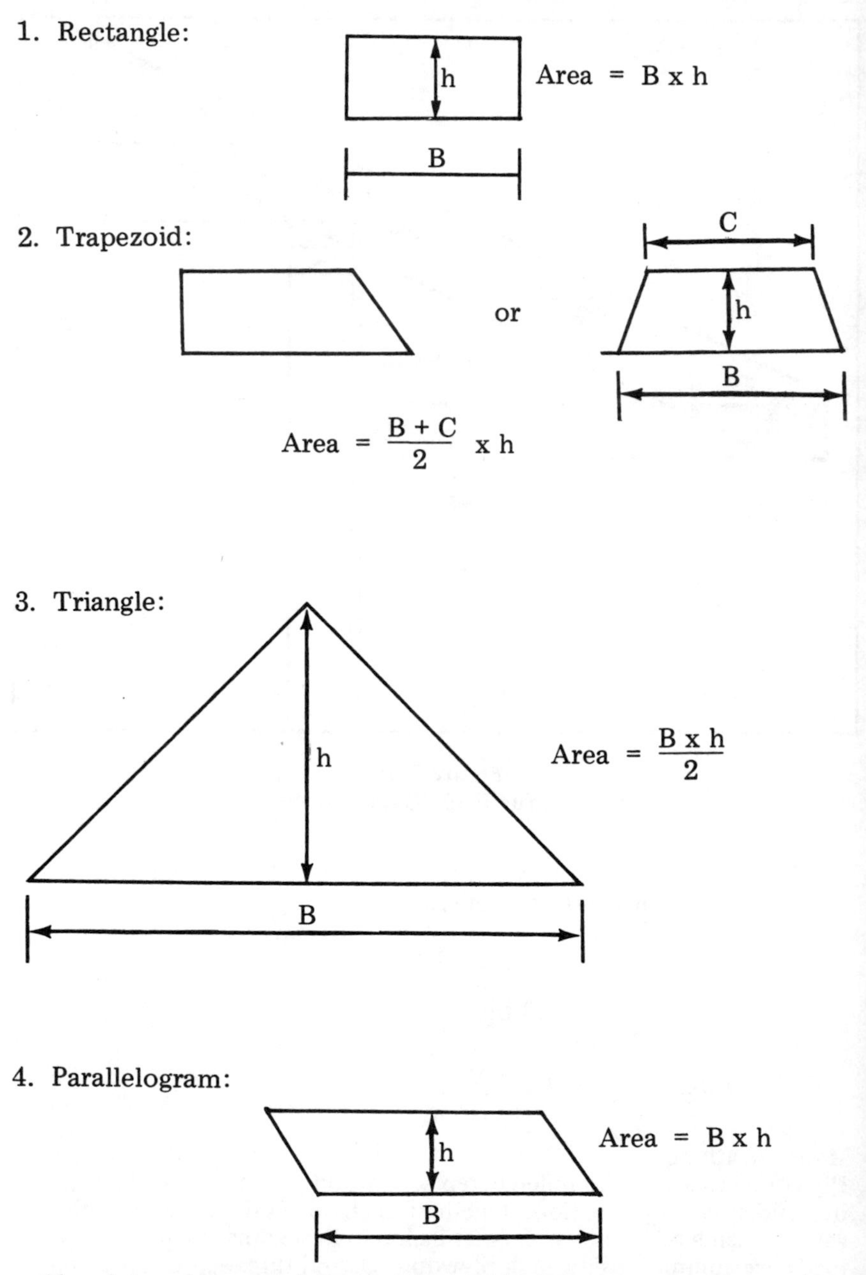

The plain gable roof shown in Figure 7-30 is a rectangle. The formula for computing the area of a rectangle is:

Area = B x h

"B" is the length of the eave A-B and "h" is the length of the rafter to the center of the ridge. The roof area in Figure 7-30 is computed as follows:

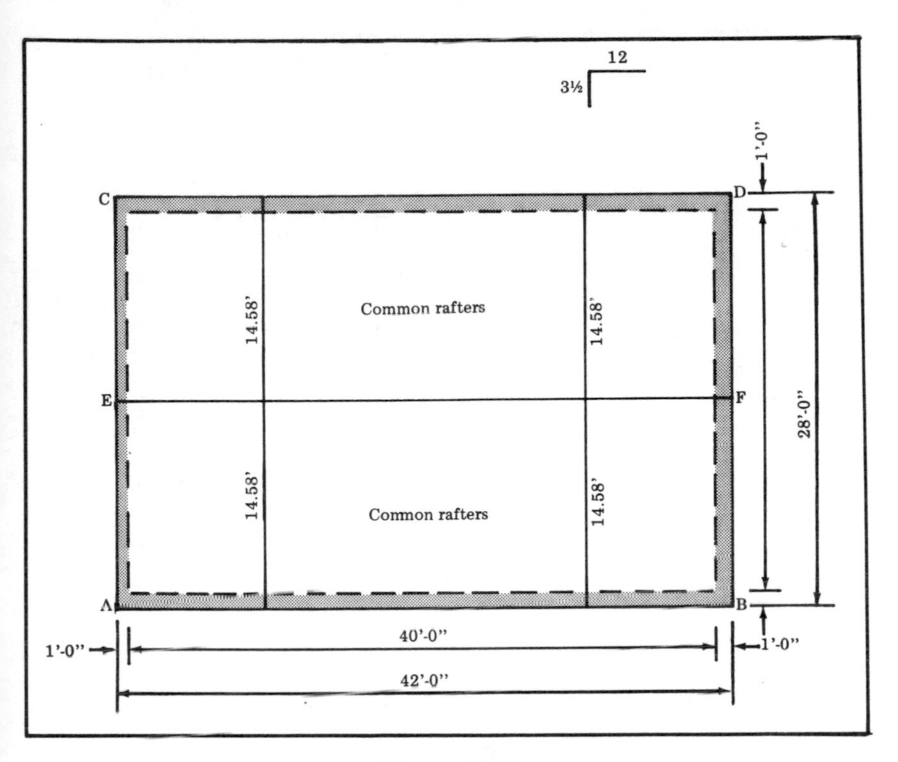

Figure 7-30
Gable Roof Area

Area = A-B x rafter length
Area = 42.0' x 14.58' (*)
Area = 612.36 square feet (for section A-B-E-F)

The roof section C-D-E-F is the same size as A-B-E-F so the total roof area in Figure 7-30 is:

612.36 sq. ft. x 2 = 1224.72 sq. ft.

(*) Note: The length of the rafters is shown in feet and decimal equivalents to make calculations easier. To convert feet and decimal equivalents to feet and inches, use the conversion chart provided (Figure 7-21).

The plain hip roof in Figure 7-31 is both a trapezoid (sections A-B-E-F and C-D-E-F) and a triangle (sections B-F-D and C-E-A). The formula for finding the area of a trapezoid is:

$$\text{Area} = \frac{B + C}{2} \times h$$

"B" is the length of the eave A-B or C-D, "C" is the length of the ridge E-F and "h" is the length of the rafter to the center of the ridge. The roof area in Figure 7-31 is:

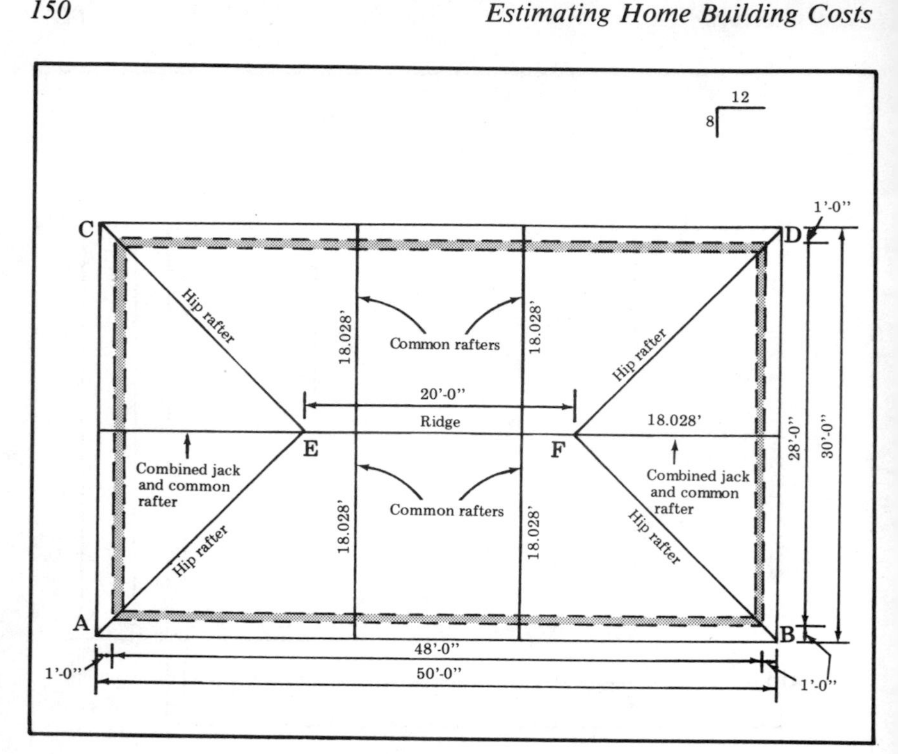

Figure 7-31
Hip Roof Area

$$\text{Area} = \frac{\text{A-B} + \text{E-F}}{2} \times \text{rafter length}$$

$$\text{Area} = \frac{50.0' + 20'\text{-}0''}{2} \times 18.028'$$

Area = 630.98 sq. ft. (For section A-B-E-F)
630.98 x 2 = 1261.96 sq. ft. (For both sections)

The formula for the triangle is:

$$\text{Area} = \frac{B \times h}{2}$$

"B" is the length of the eave B-D or C-A and "h" is the length of the combined jack and common rafter. The formula for the triangle roof in Figure 7-31 is:

$$\text{Area} = \frac{\text{B-D} \times \text{rafter length}}{2}$$

$$\text{Area} = \frac{30.0' \times 18.028'}{2} = 270.42 \text{ sq. ft.}$$

270.42 x 2 = 540.84 sq. ft. (Sections B-F-D and C-E-A)

The total roof area in Figure 7-31 is:

Two trapezoid sections	1261.96 sq. ft.
Two triangle sections	540.84 sq. ft.
Total	1802.80 sq. ft.

Note: If the plain hip roof in Figure 7-31 was calculated as a rectangle, the computations would be:

Area = A-B x length of rafter
Area = 50 x 18.028'
Area = 901.40 sq. ft.
(*) 901.40 x 2 = 1802.80 sq. ft. (Both sections)

(*) This is the same area as the separate computations for the trapezoids and triangles. *Conclusion: the roof area for a plain hip roof can be computed as a rectangle.*

To estimate the roof area of an offset or intersection with a gable roof (Figure 7-32):

1. Compute the main roof of section A-B-E-F as solid; this will include the area of the triangle I-J-B in the offset.
2. Multiply the eave length I-G by the common rafter length in the offset for the area of one section. Multiply the area of one section by 2 for the total area. Example:

Area = I-G x rafter length
Area = 24.0' x 12.30' = 295.20 sq. ft.
295.20 x 2 = 590.40 sq. ft.

The roof area for the offset with a gable roof in Figure 7-32 is:
590.40 square feet (*)
(*) Add this roof area to the roof area of the main roof.

To estimate the roof area of an offset or intersection with a hip roof (Figure 7-32):

1. Compute the main roof area of section A-B-E-F as solid; this will include the area of the triangle I-J-B in the offset.
2. The remaining roof section I-G-K-L is a trapezoid and the roof area is:

$$\text{Area} = \frac{\text{I-G} + \text{K-L}}{2} \times \text{rafter length}$$

$$\text{Area} = \frac{24.0' + 13.0' \text{ (not shown)}}{2} \times 12.30'$$

$$\text{Area} = \frac{37}{2} \times 12.30 = 227.55 \text{ sq. ft.}$$

227.55 sq. ft. x 2 = 455.10 sq. ft.

The area of the triangle I-J-B is: $\frac{22 \times 12.30}{2} = 135.30$ sq. ft.

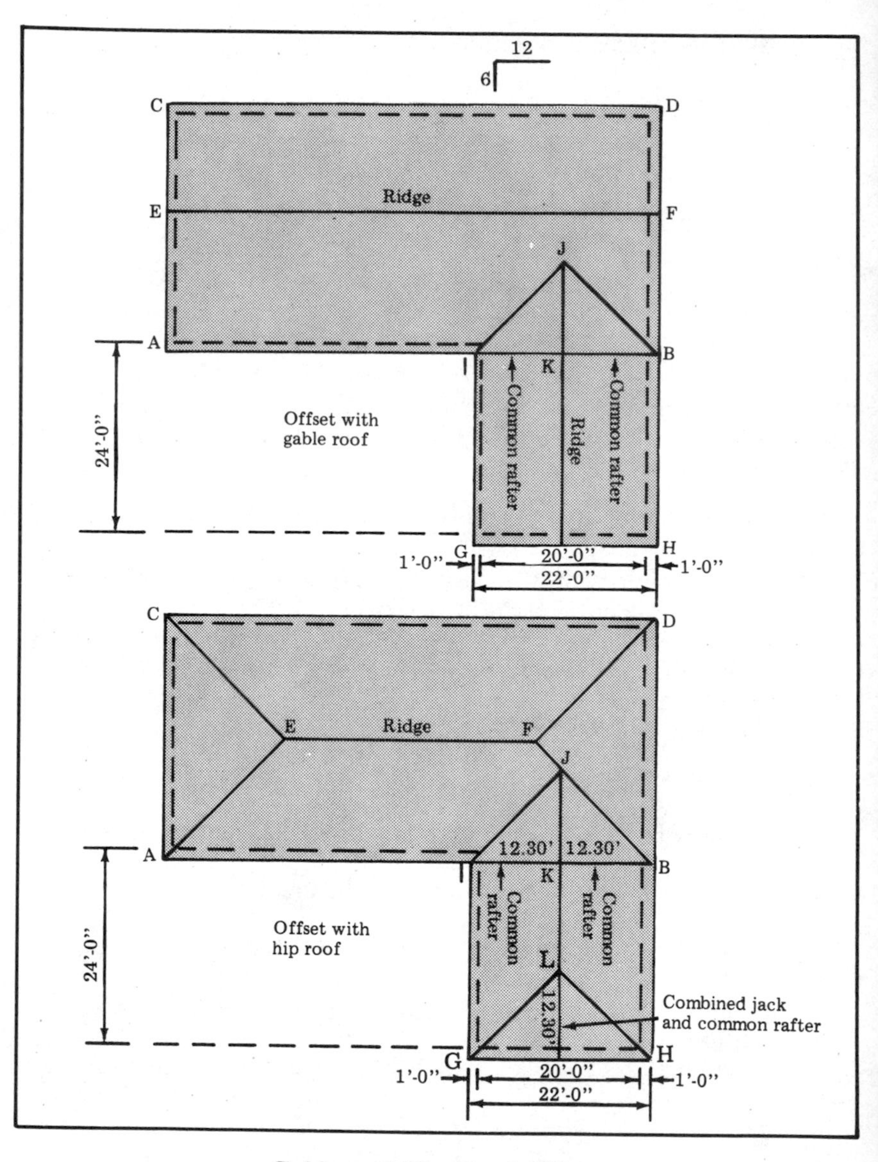

Gable and Hip Roof Offset
Figure 7-32

The roof area for the offset with a hip roof in Figure 7-32 is:

2 — Trapezoids	455.10 sq. ft.
1 — Triangle	135.30 sq. ft.
Total	590.40 sq. ft.

Note: The roof area of the offset with a gable roof (590.40 square feet) is the same as the offset with the hip roof. *Therefore, it is not necessary to calculate the trapezoid and triangle sections separately. The offset with the hip roof can be calculated as a rectangle.*

The length of the rafters in Figures 7-30, 7-31 and 7-32 is shown in feet and decimal equivalents for easier calculations.

A plain hip roof running to a point consists of four triangles. To get the roof area of the four sections, calculate the area of one section and multiply by 4.

A faster method of computing roof areas is to multiply the ratios for the roof pitch shown in Figure 7-24 by the square feet of the floor area plus the roof extension. Example: in Figure 7-30 the roof pitch is 3½/12, and the area including the roof overhang is 1176 square feet (42.0' x 28.0'). The ratio for a 3½/12 roof pitch from Figure 7-24 is 1.042.

1176 sq. ft. x 1.042 = *1225.39 sq. ft.*

The previous calculations for the roof area in Figure 7-30 were *1224.72 square feet.*

Figure 7-31 is a plain hip roof with a pitch of 8/12. The area, including the roof overhang, is 1500 square feet (50.0' x 30.0'). The ratio for a 8/12 roof pitch from Figure 7-24 is 1.202.

1500 sq. ft. x 1.202 = *1803.00 sq. ft.*

The previous calculations for the hip roof area were 1802.80 square feet.

Note: If you are using these ratios to compute roof areas, be especially careful in the offsets or intersections. Any break in the roof line can result in erroneous computations.

The roof area for the house in Figures 7-1, 7-2 and 7-10 is:

First floor	729.00 sq. ft.
Second floor	1224.72 sq. ft.
Total	1953.72 sq. ft.

The roof sheathing will be ½" x 4' x 8' plywood. The number of pieces that will be required is:

$$\frac{1953.72}{32} = 61.05 \text{ or } 62 \text{ pieces}$$

Order: *62 pieces ½" x 4' x 8' plywood (1984 sq. ft.)*

Nails

The quantity of nails for the ceiling and roof framing, and the sheathing is computed as follows:

Framing allowance:	8d common nails	2 lbs. per 1000 bd. ft.
	16d common nails	10 lbs. per 1000 bd. ft.
Plywood allowance:	6d threaded nails	12 lbs. per 1000 sq. ft.

There are 3659 board feet of framing lumber in the roof system and the quantity of nails that will be ordered is:

8d common nails

$$\frac{3659 \times 2}{1000} = 7.32 \text{ or } \textit{8 lbs.}$$

16d common nails

$$\frac{3659 \times 10}{1000} = 36.59 \text{ or } \textit{37 lbs.}$$

There are 1984 square feet of plywood for the roof sheathing, so the nail requirement will be:

6d threaded nails

$$\frac{1984 \times 12}{1000} = 23.81 \text{ or } \textit{24 lbs.}$$

The nail quantity for the roof system will be:

8 lbs. 8d common nails
37 lbs. 16d common nails
24 lbs. 6d threaded nails

Trusses

Roof trusses are a framework of individual structural members fabricated into one unit and designed to span a large distance; interior partitions are unnecessary. They should be designed by a qualified engineer or architect in accordance with standard engineering practice.

Trusses are available in a variety of sizes and shapes and can be used for long or short spans such as the roof over a porch, walk, or breezeway. They are normally spaced 24 inch o.c. and save labor and material costs. Trusses eliminate the following framing in the roof system:

1. Ceiling joists
2. Rafters
3. Ridge
4. Collar beams
5. Rafter supports
6. Lookouts
7. Gable end studs

The bottom chord of the truss is used as a ceiling joist, and the top chord is used as the rafter. The roof overhang at the eave can be fabricated with the truss as shown in Figure 7-33. Blocking for the gable roof is cut from 2 x 6 or 2 x 8 framing and used in the overhang at 16 inch o.c. or 24 inch o.c. They are fabricated in sections about 10 feet in length and fastened to the top chord of the truss with metal straps (Figure 7-34).

To estimate the number of roof trusses on 24 inch o.c., multiply the length of the wall perpendicular to the truss by .50 and add one. Do this for each section of the house where there is a break in the roof line. Example: the number of trusses required for the first and second floors for the house in Figures 7-1 and 7-2 is:

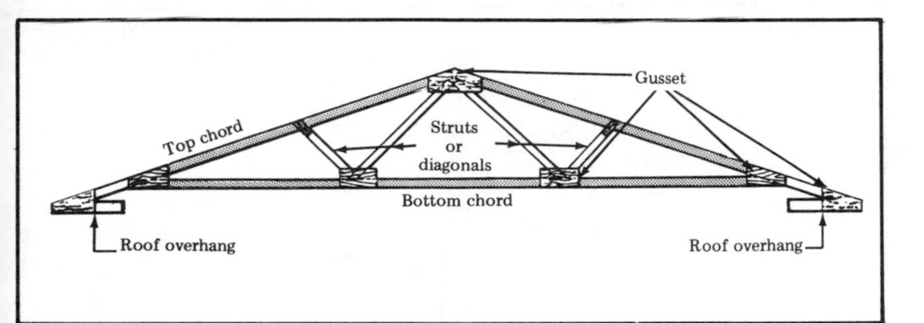

Figure 7-33
Truss with Roof Overhang

Second floor 40.0' x .50 = 20
20 + 1 = *21 trusses*

First floor 24.0' x .50 = 12
12 + 1 = *13 trusses*

Order: *34 - trusses* with fabricated blocking for the gable overhang.

Nail the trusses with two 10d common nails on each side of the truss at each wall plate. Obviously the nail allowance for trusses will be much less than for ceiling joists and rafters.

Order: 10 lbs. 10d common nails

(Enter all material that is to be used for the ceiling and roof framing on Line 7.2 on the Cost Estimate Worksheet)

Porch Shed Roof Framing

When shed roofs are built to cover porches and walks (Figures 7-35 and 7-37) and they are attached to the house, they are framed as shown in Figure 7-36. Either ceiling joists and rafters, or trusses, can be used.

If rafters are used, estimate the number of ceiling joists and rafters in the same manner as was done for the main house. The length of the rafters is computed in the same manner as the length of the common rafters was with one exception. For the common rafters of a house, one-half of the thickness of the ridge is deducted from the length of the rafter for the "true rafter length"; whereas for the porch shed roof, the full thickness of the rafter nailer (serves the same purpose as the ridge) is deducted for the "true rafter length" (Figure 7-36 and the worksheet Figure 7-36A). Sample worksheets are shown.

If ceiling joists and rafters were used for the shed roof covering the terrace in Figure 7-35 (The roof overhang is 6 inches for each gable and the eave) the material list will be:

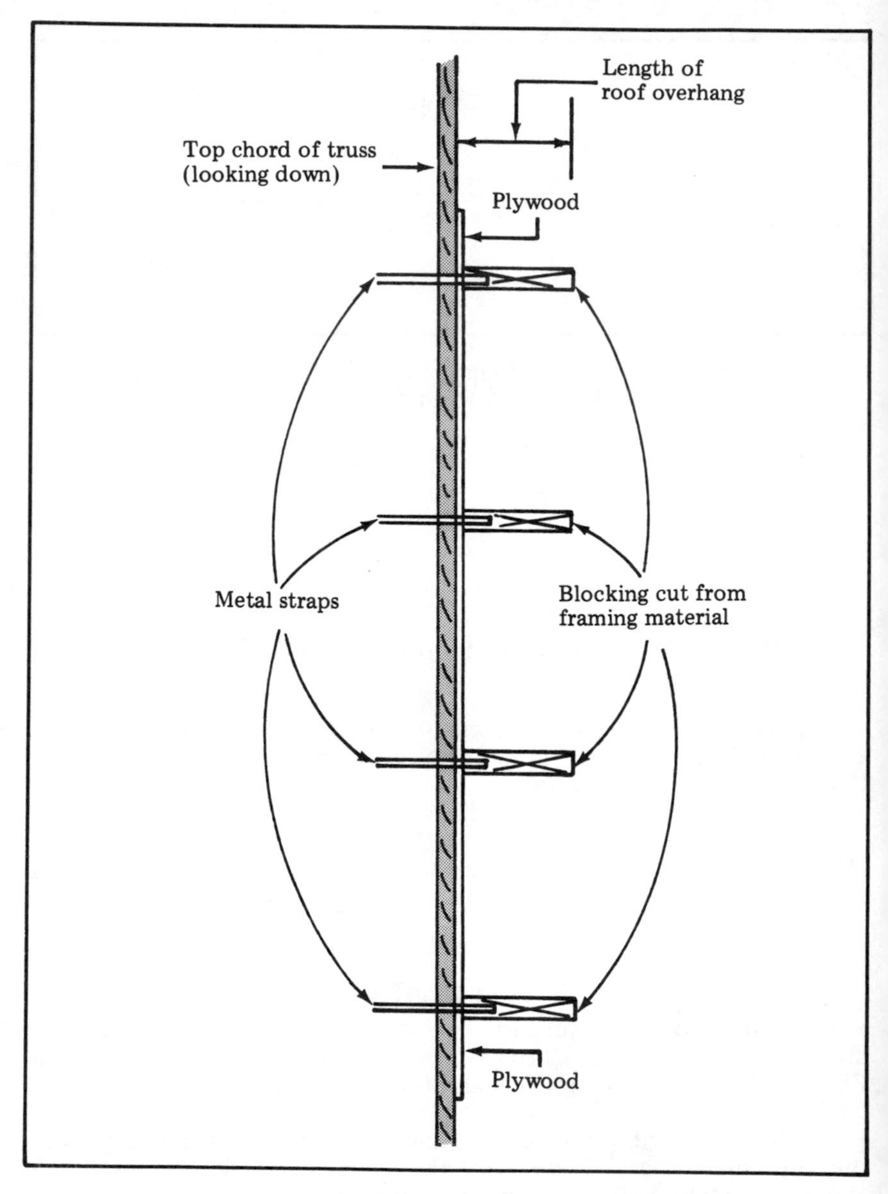

Figure 7-34
Fabricated Gable Roof Overhang

Built-Up-Beam: . . . . .4 — 2 x 6 x 12 (48 bd. ft.)
2 — 2 x 6 x 18 (36 bd. ft.)

Nailer For Ceiling Joists: . . . . .1 — 2 x 6 x 18 (18 bd. ft.)

Ceiling Joists: (16" o.c.) . . . . .14 — 2 x 6 x 12 (168 bd. ft.)

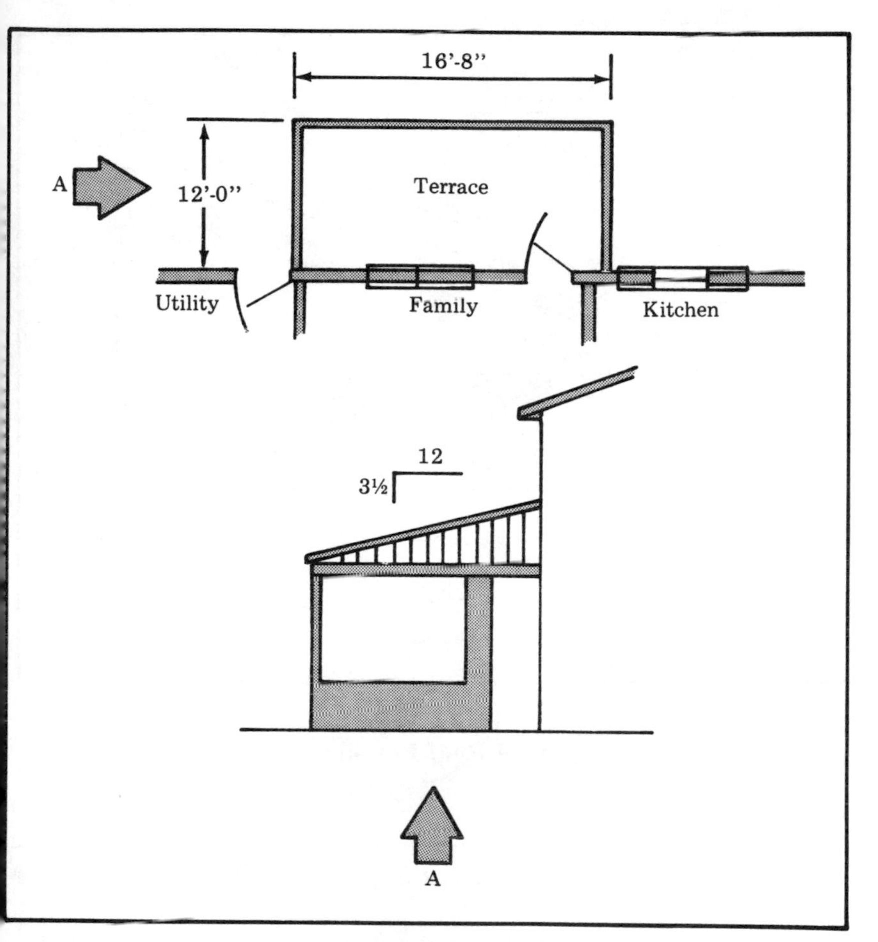

Figure 7-35
Porch Shed Roof

Rafter nailer: 1 - 2 x 8 x 18 (24 bd. ft.)
Rafters: 15 - 2 x 6 x 14 (210 bd. ft.)
Total (504 bd. ft.) includes totals from page 156.
Roof sheathing: 8 pcs. ½" x 4' x 8' plywood (256 sq. ft.)
Nails: 7 lbs. 16d common nails
3 lbs. 6d threaded nails

Trusses were used for the shed roofs in Figure 7-35 and 7-37, and are on 24 inch centers. The material list for the trusses will be:

Terrace (Figure 7-35)
Built-up-beam: 4 - 2 x 6 x 12 (48 bd. ft.)
2 - 2 x 6 x 18 (36 bd. ft.)
Truss nailer: 2 - 2 x 4 x 18 (24 bd. ft.)

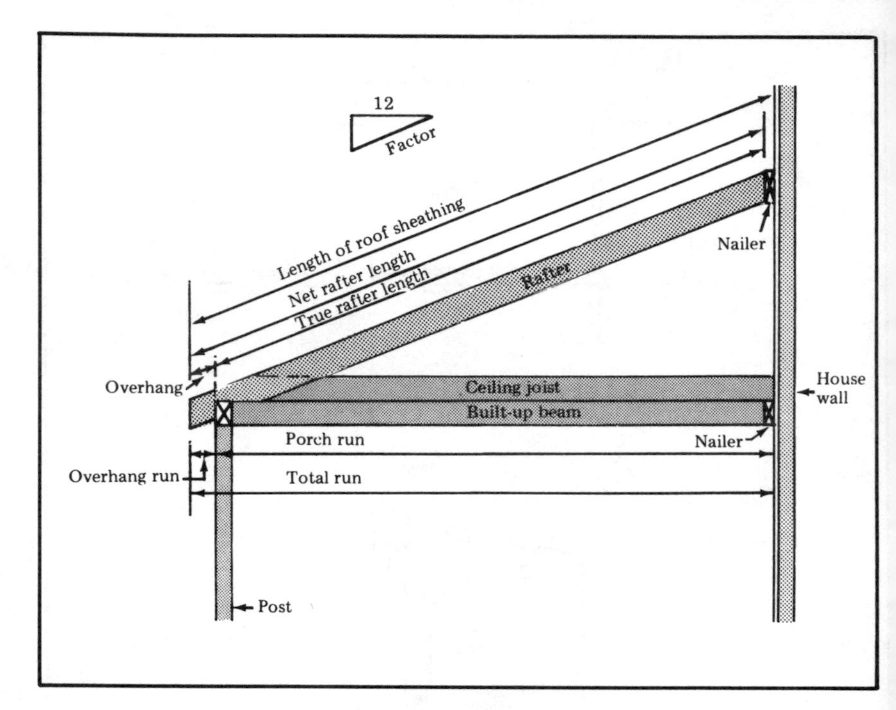

Figure 7-36
Shed Roof Framing

Trusses:	10
Roof sheathing:	8 pcs. ½" x 4' x 8' plywood (256 sq. ft.)

Front Porch and Walk (Figure 7-37)

Built-up-beam:	1 - 2 x 6 x 18 (18 bd. ft.)
	2 - 2 x 6 x 16 (32 bd. ft.)
	4 - 2 x 6 x 12 (48 bd. ft.)
	1 - 2 x 6 x 8 (8 bd. ft.)
Truss nailer:	1 - 2 x 4 x 16 (11 bd. ft.)
	2 - 2 x 4 x 12 (16 bd. ft.)
Trusses:	22
Roof sheathing	13 pcs. ½" x 4' x 8' plywood (416 sq. ft.)
Nails	6 lbs. 16d common nails
	5 lbs. 6d threaded nails

(Enter on Line 7.3 Cost Estimate Worksheet)

Stair Stringers

If there is more than one floor level in a house, it is wise to install the stair stringers and temporary treads between floors as soon as possible

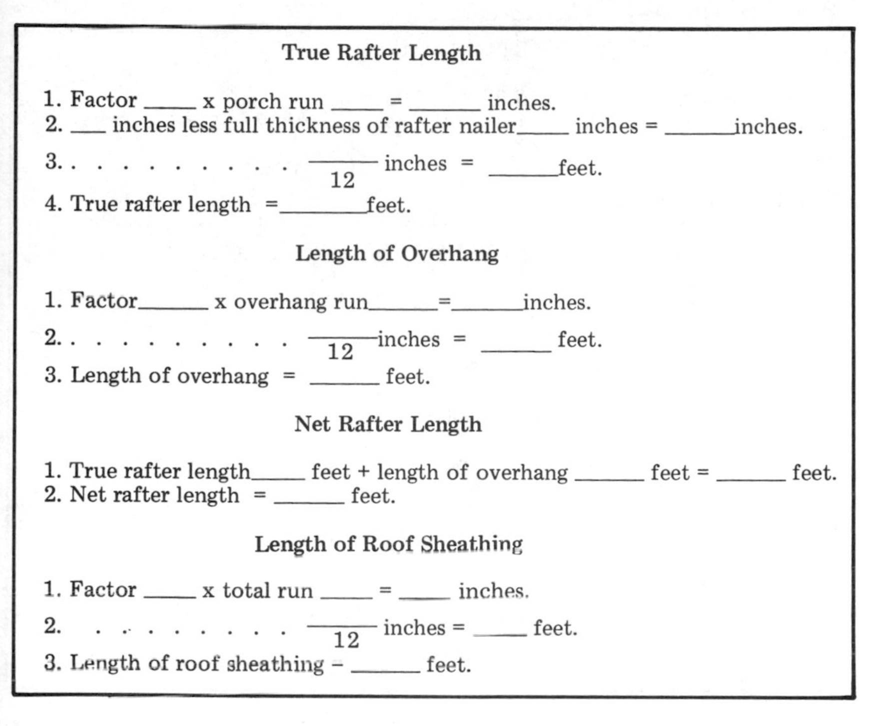

True Rafter Length

1. Factor ____ x porch run ____ = ______ inches.
2. ___ inches less full thickness of rafter nailer____ inches = ______inches.
3. . . . . $\frac{\_\_\_\_}{12}$ inches = ______feet.
4. True rafter length =________feet.

Length of Overhang

1. Factor______ x overhang run______=______inches.
2. . . . . $\frac{\_\_\_\_}{12}$ inches = ______ feet.
3. Length of overhang = ______ feet.

Net Rafter Length

1. True rafter length_____ feet + length of overhang ______ feet = ______ feet.
2. Net rafter length = ______ feet.

Length of Roof Sheathing

1. Factor ____ x total run ____ = ____ inches.
2. . . . . $\frac{\_\_\_\_}{12}$ inches = ____ feet.
3. Length of roof sheathing = ______ feet.

Figure 7-36A (Worksheet)

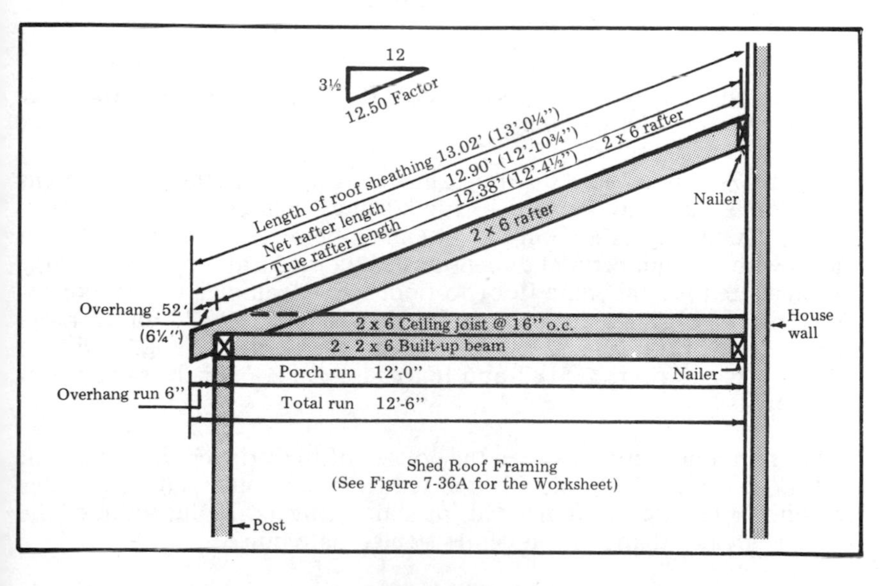

Figure 7-36
Shed Roof Framing
Sample

True Rafter Length

1. Factor 12.50 x porch run 12.0' = 150 inches.
2. 150 inches less full thickness of rafter nailer 1.5 inches = 148.5 inches.
3. . . . . $\frac{148.5}{12}$ inches = 12.38 feet.
4. True rafter length = 12.38 feet.

Length of Overhang

1. Factor 12.50 x overhang run .5' (6") = 6.25 inches.
2. . . . . $\frac{6.25}{12}$ inches = .52 feet.
3. Length of overhang = .52 feet.

Net Rafter Length

1. True rafter length 12.38 feet + length of overhang .52 feet = 12.90 feet.
2. Net rafter length = 12.90 feet.

Length of Roof Sheathing

1. Factor 12.50 x total run 12.5' = 156.25 inches.
2. . . . . $\frac{156.25}{12}$ inches = 13.02 feet.
3. Length of roof sheathing = 13.02 feet.

Figure 7-36A
(Worksheet)
Sample

after the house is under temporary roof. It is more convenient and safer for the workmen on the job.

Figure 7-38 is a table of approximate stringer lengths for different floor levels that may be used as a guideline when estimating stair stringers. These lengths are only approximate because the stringer length varies with the number and dimensions of the risers and treads that may be calculated for the same floor-to-floor rise. Example: for a floor-to-floor rise of 8'10⅝", the stringer length will be 12'4$^{13}/_{16}$" for 13 risers @8$^{3}/_{16}$" and 12 treads @9$^{5}/_{16}$"; whereas the stringer length will be 17'1$^{29}/_{32}$" for 17 risers @6¼" and 16 treads @11¼" for the same floor-to-floor rise.

The minimum depth between the bottom of the stringer and the cutout is 3½" (Figure 7-39), unless supported by other construction. For this reason 2 x 12s are recommended for stair stringers. If the width of the stairs is greater than 2'6", a center stringer is required.

There is normally enough waste material from the framing and plywood without ordering additional material for the temporary treads.

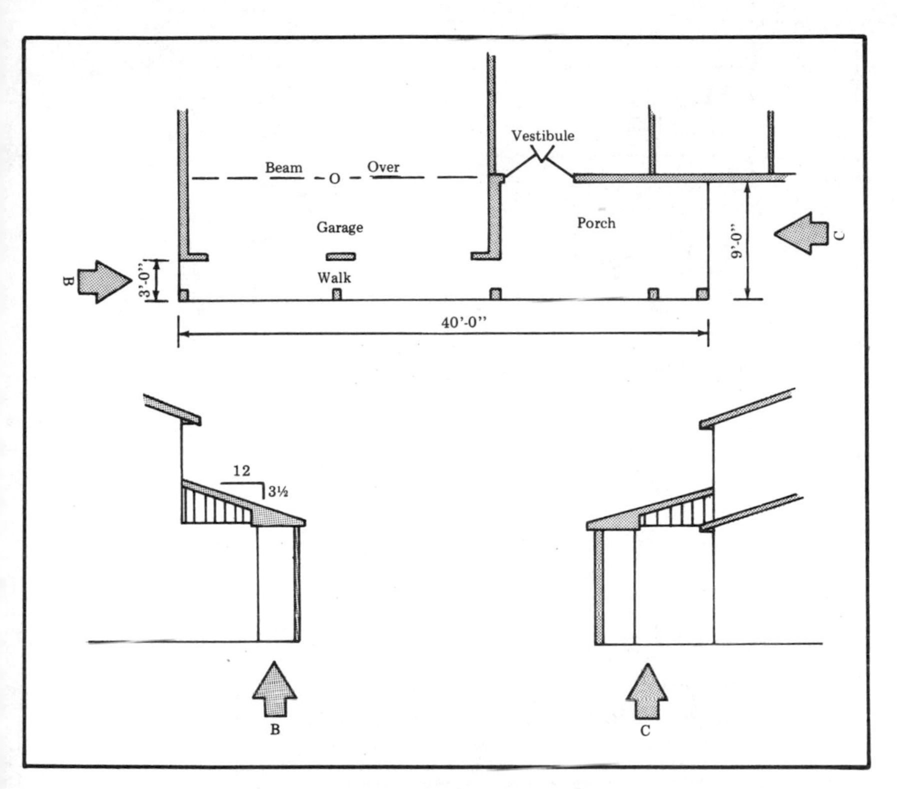

Porch and Walk Shed Roof
Figure 7-37

Most builders do not construct the basement stairs until after the concrete basement floor is poured. The basement stair stringers in this estimate will be estimated later in this book.

The material required for the stair stringers from the first to the second floor in Figures 7-1, 7-2 and 7-39 is:

Order: 3 - 2 x 12 x 14 (84 bd. ft.)

(Enter on Line 7.4 Cost Estimate Worksheet)

Factory-Built Houses

Factory-built houses, as explained here, are sectional and not modular houses. Modular houses are completely finished at the factory and transported to the job site in units of two or more sections and set on the builder's foundation. Sectional houses are fabricated at the factory in sections small enough for four or more workmen to handle. The exterior wall sections will be assembled in panels, probably not exceeding 16'0"

Floor to Floor Rise	Approximate Length of Stringers (*)
3'-0"	6'-0"
4'-0"	7'-0"
4'-6"	7'-6"
5'-0"	9'-0"
5'-6"	9'-6"
6'-0"	10'-0"
6'-6"	11'-0"
7'-0"	12'-0"
7'-6"	12'-6"
8'-0"	13'-0"
8'-6"	14'-0"
9'-0"	15'-0"
9'-6"	15'-6"
10'-0"	16'-0"

(*) The stair stringer length varies with the same floor-to-floor rise depending on the number and dimensions of the risers and treads. The length of the stringer increases as the number of risers increases.

Approximate Stringer Lengths
Figure 7-38

in length for the first floor, 14'-0" for the second floor. The wall sheathing and windows (excluding large picture windows) will be in place. All door openings will be framed in; the upper top plate will be shipped loose.

The basic package will include all exterior and interior walls, nails, prehung doors including locks, windows including locks and pulls, and the floor system for the second floor (the floor system for the first floor is an option unless it is a special design; then it will probably be included in the basic package). The roof system is included in the basic package and trusses will be included when the roof pitch and design permit their use. Rafters and ceiling joists will be included when trusses are not. The roof system will include the roof sheathing.

The basic package for the interior and exterior trim will include everything required for the house, from nails to porch beams to closet shelves. Exterior siding, kitchen and bathroom cabinets, garage doors, appliances, and roofing are among the many options that may be ordered with the house package.

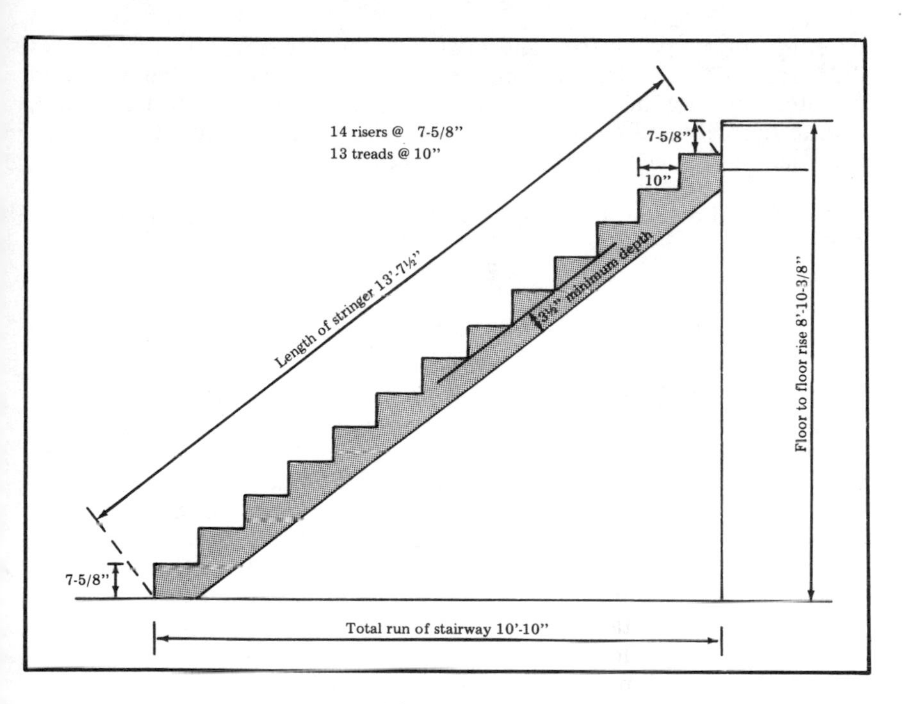

Figure 7-39
Stair Layout

Quality materials and workmanship are a prerequisite in most factory-built houses. Some of the advantages of using factory-built houses are:

1. No fee for the blueprints.
2. Faster build-out time.
3. Less on-site labor costs.
4. Less estimating time.
5. Better cost controls.

Some of the disadvantages are:

1. High freight costs delivering the package house to the job site.
2. Increased labor costs due to truck delays.
3. If the factory is a considerable distance from the job site there may be delays in replacing damaged materials and in procuring additional supplies.

The two-story house with a double garage shown in Figures 7-1 and 7-2 is a factory-built house. The basic package includes all of the material listed above including the floor system for the first floor (the first floor had special trusses), laminated beams, and the shed roof system, including trusses, porch trim and posts for the porches shown in Figures 7-35 and 7-36. The optional material will be the kitchen cabinets and counter tops, bathroom cabinets and counter tops, exterior siding, prefab fireplace and all chimney material. All of this optional material is included in the package price of the house shown below.

The cost of the house package including freight to the job site and sales tax is:

$21,087.84

(Enter on Line 7.5 Cost Estimate Worksheet)

Labor Costs For The Superstructure

Labor costs are always an estimate and too much emphasis cannot be put on accurate cost records from previous jobs for estimating the labor for any job. Another factor that affects labor costs is weather. The construction of the superstructure of a house is outside work and an allowance should be made for lower productivity in cold weather. Depending on the severity and duration of the cold weather, an additional 5%-10% should be added to the estimated man-hours for cold-weather construction.

The labor costs will be estimated from costs for similarly built houses recently constructed.

Custom-Built With Ceiling Joists And Rafters

Payroll records show that the superstructure of a two-story house with 2183 square feet including the garage required 507 man-hours. This house was similar in design to the house in Figures 7-1, 7-2 and 7-10, but was custom-built using ceiling joists and rafters. The house shown in the above drawings has 2879 square feet, including the garage. The man-hours to construct the superstructure of this house will be estimated as follows:

$$507:2183 = x:2879$$

$$507:x = 2183:2879$$

$$x = 668.65 \text{ or } 669 \text{ man-hours}$$

The time estimated for the superstructure for this custom-built house will be:

669 Man-Hours

For the estimator who may not have past labor cost records to use as a guideline, the following labor factor can be used for estimating the labor for the superstructure of a custom-built house using ceiling joists and rafters:

Man-Hour Factor: .23225 per square foot
Example: .23225 x 2879 square feet = 668.65 or 669 man-hours
Example: .23225 x 2183 square feet = 507 man-hours
Note: Trusses where usable require approximately one-third fewer man-hours for the roof system than ceiling joists and rafters.

Factory-Built House Using Trusses

The payroll records for a similarly designed factory-built house with 2286 square feet including the garage show that 321 man-hours were required for the superstructure. Using these records, the man-hours for the factory-built house in Figures 7-1, 7-2, and 7-10 (2879 square feet including the garage) will be estimated at:

$$321:2286 = x:2879$$
$$321:x = 2286:2879$$
$$x = 404.27 \text{ or } 404 \text{ manhours}$$

The time estimated for the superstructure of the factory-built house will be:

(*) 404 Man-Hours

The man-hour factor for the superstructure of a factory-built house using trusses is:

Man-Hour Factor: .14042 per square foot
Example: .14042 x 2879 square feet = 404.27 or 404 man-hours
Example: .14042 x 2286 square feet = 321 man-hours

(*)Note: This is an estimated saving of 265 man-hours, or approximately 40% saving in labor from the estimated 669 man-hours for the custom-built house using ceiling joists and rafters.

Six workmen will be assigned to this job and their pay scale is shown in Chapter 6. The man-hours assigned to each man will be:

$$\frac{404}{6} = 67.33 \text{ or } 68 \text{ man-hours}$$

The estimated labor cost for the superstructure will be:

68 hours	@	$8.50 x 1	=	$578.00
68 hours	@	6.25 x 2	=	850.00
68 hours	@	5.00 x 3	=	1020.00
		Total		$2448.00

(Enter this labor estimate on Line 7.6 Cost Estimate Worksheet)

Cost Estimate Worksheet For Superstructure

7.1 Exterior and Interior Walls:

Sole and top plates

_____pcs. (size______) @______= $_____

Studs

_____pcs. (size______) @______= _____

Headers

_____pcs. (size______) @______= _____
_____pcs. (size______) @______= _____
_____pcs. (size______) @______= _____
_____pcs. (size______) @______= _____
_____pcs. (size______) @______= _____
_____pcs. (size______) @______= _____
_____pcs. (size______) @______= _____
_____pcs. (size______) @______= _____
_____pcs. (size______) @______= _____
_____pcs. (size______) @______= _____

Temporary wall braces

_____pcs. (size______) @______= _____

Corner bracing

_____pcs. (size______) @______= _____

Wall sheathing

_____pcs. (size______) @______= _____
(type______)
_____pcs. (size______) @______= _____

_____pcs (type_____)

Nails

_____lbs. (size_____) @______=______
_____lbs. (size_____) @______=______
_____lbs. (size_____) @______=______

Line 7.1 $_______ $_________

7.2 Ceiling and Roof Framing:

Ceiling joists

_____pcs. (size______) @______= $______
_____pcs. (size______) @______= ______
_____pcs. (size______) @______= ______
_____pcs. (size______) @______= ______

Ceiling backing

_____pcs. (size______) @______= ______

Ceiling joist stiffeners
_____lin. ft. (size______) @______ = ______

Rafters
_____pcs. (size______) @______ = ______
_____pcs. (size______) @______ = ______
_____pcs. (size______) @______ = ______
_____pcs. (size______) @______ = ______

Ridge
_____pcs. (size______) @______ = ______

Collar beam
_____pcs. (size______) @______ = ______

Rafter supports
_____pcs. (size______) @______ = ______

Lookouts
_____pcs. (size______) @______ = ______

Roof sheathing
_____pcs. (size______) @______ = ______
(type_____)

Nails
_____lbs. (size______) @______ = ______
_____lbs. (size______) @______ = ______
_____lbs. (size______) @______ = ______

Trusses
_____units @______ = ______ ________
_____units @______ = ______ ________

Nails
_____lbs. (size______) @______ = ______

Line 7.2 $________ $________

7.3 Porch Shed Roof Framing and Sheathing: (with ceiling joists and rafters)

Built-up beam (or porch rim)
_____pcs. (size______) @______ = $______
_____pcs. (size______) @______ = ______
_____pcs. (size______) @______ = ______

Nailer for ceiling joists
_____pcs. (size______) @______ = ______

Ceiling joists (spacing o.c.)
_____pcs. (size______) @______ = ______

Rafter nailer (or ridge)
_____pcs. (size______) @______ = ______

Rafters
_____pcs. (size______) @______ = ______
_____pcs. (size______) @______ = ______

Roof sheathing
_____pcs. (size______) @______ = ______
(type_____)

Nails
_____lbs. (size______) @______ = ______
_____lbs. (size______) @______ = ______

Built-up beam (or porch rim)
_____pcs. (size______) @______ = ______
_____pcs. (size______) @______ = ______
_____pcs. (size______) @______ = ______

Trusses
_____units @______ = ______ ______
_____units @______ = ______ ______

Roof sheathing
_____pcs. (size______) @______ = ______
_____pcs. (size______) @______ = ______
(type_____)

Nails
_____lbs. (size______) @______ = ______
_____lbs. (size______) @______ = ______

Line 7.3 $______ $________

7.4 **Stair stringers:**
_____pcs. (size______) @______ = $______
_____pcs. (size______) @______ = ______

Line 7.4 $______ $________

Cost of material (add lines from 7.1, 7.2, 7.3, and 7.4) $________

Sales tax (____ %) ________

Total cost of material $________ (1)

7.5 **Factory-Built Houses:**
Total cost of package delivered to job site including sales tax and freight $________ (2)

7.6 **Labor Costs for Superstructure** $________ (3)

Total Cost:

(a) Custom-built house (add lines (1) and (3) $ ________

(b) Factory-built house (add lines (2) and (3) $________

(Enter either line (a) or line (b) on Line 7, Form 100)

Chapter 8

Roofing

Roofing is the covering over the roof deck that prevents the entrance of moisture. In residential construction asphalt and fiberglass shingles account for approximately 75% - 85% of the material used, followed by wood shingles and built-up roofing. The roofing materials that will be included in this chapter are:

1. Felt
2. Flashing and plastic roof cement
3. Metal Drip Edge
4. Nails

Regardless of the quality of the roofing material, if the felt, or underlayment, and the flashing are not properly applied, and the roofing is not nailed down according to the manufacturer's recommendation, the roof will not be waterproof.

A good roof is a wise investment and the initial cost of the material should not be the deciding factor when selecting the roofing. Roofing material varies in cost and while one may be cheaper to install initially, another type that is more expensive may prove cheaper over the years. In selecting the roof material, the builder and the owner must take into consideration many factors including the type of building, the roof pitch and local weather conditions.

Under-layment	Minimum Roof Slope	
	Double Coverage Shingle *	Triple Coverage Shingle *
Not required	7 in 12	4 in 12
Single	4 in 12	3 in 12
Double	2 in 12	2 in 12

*Double coverage for a 12" x 36" shingle is considered to be an exposure of 5". Triple coverage is considered to be an exposure of 4".
Note: The headlap for single coverage of underlayment should be 2" and for double coverage 19".

Underlayment for Asphalt and Fiberglass Shingles
Figure 8-1

Roof Covering

Felt:
Roofing felt is an asphalt impregnated rag fiber paper underlayment manufactured in rolls. Felt weighing approximately 15 lbs. per square comes in rolls 3'0" wide x 144'0" long (432 square feet). The felt should be provided over the entire roof on those roof pitches shown in Figure 8-1.

To estimate roofing felt, divide the roof area by 400 and round off to the next highest number for 15 lb. weight. *Example:*

$$\frac{1834 \text{ sq. ft.}}{400} = 4.59 \text{ or } 5 \text{ rolls}$$

For 30 lb. felt divide the roof area by 200 and round off to the next highest number. *Example:*

$$\frac{1834 \text{ sq. ft.}}{200} = 9.17 \text{ or } 10 \text{ rolls}$$

Flashing
Flashing provides watertightness at all critical joints in roof construction. It is installed in valleys and all intersections at walls and chimneys. Counter flashing is a second and overlapping layer of flashing used as an additional safeguard.

It is used primarily at the intersection between the roof and masonry walls and around chimneys and chimney saddles (Figure 8-2). Aluminum, asphalt, copper, galvanized sheet metal and stainless steel are acceptable flashing materials.

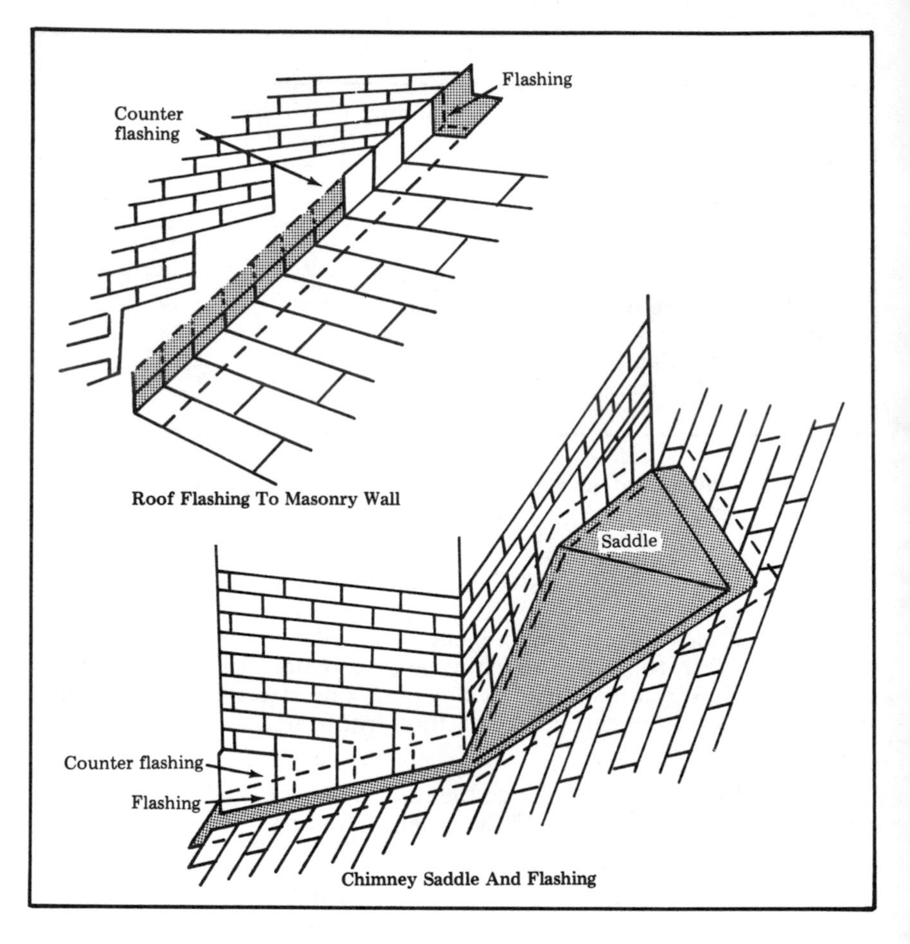

Roof and Chimney Flashing
Figure 8-2

To estimate flashing material, measure the linear footage from the elevation drawings.

Ridge Vents and Roof Ventilating Louvers:
Ridge vents (Figure 8-3) and roof ventilating louvers, when used with under eave soffit vents, provide an effective ventilating system for the attic. They can reduce summer attic temperatures as much as 50 degrees, thus reducing living quarters temperatures. When these roof vents are used, they should be installed before the roof shingles, and may be included in the estimate of roofing costs.

To estimate ridge vents, take the linear feet of the ridge, divide by 10 (10'0'' lengths) and round off to the next whole number for the number of pieces. *Example:*

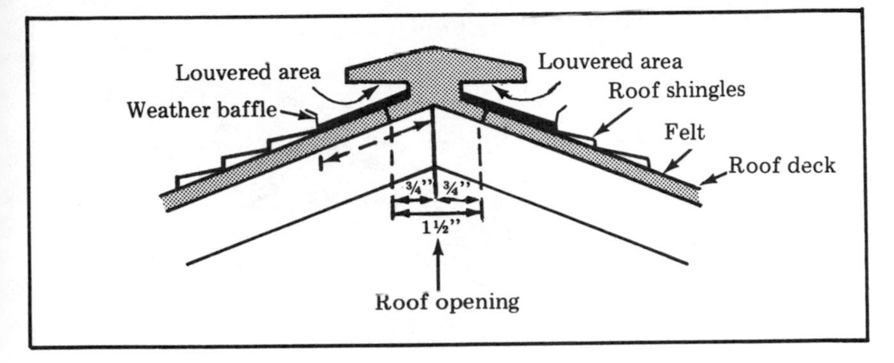

Ridge Vent
Figure 8-3

Ridge: 64 lineal feet

$$\frac{64}{10} = 6.4 \text{ or } 7 \text{ pieces}$$

Note: Add connector plugs and end plugs. Do not overlook the under eave soffit vents that should be used with the roof vents.

To estimate roof ventilating louvers, take the square feet of the attic floor and multiply by 1.2 for the square inches of net free area of ventilation required. *Example:*

Attic area: 1600 square feet
1600 x 1.2 = 1920 square inches

If 16" x 18" roof ventilating louvers will be used, divide 1920 square inches by 288 (16" x 18" = 288 square inches) for the number of ventilators required. *Example:*

$$\frac{1920}{288} = 6.67 \text{ or } 7$$

Estimate: 7 - 16" x 18" Roof ventilating louvers

10 - Under eave soffit vents

If wind driven turbine-type attic ventilators are used, follow the manufacturer's recommendation for the size and number of turbines required for the attic floor area. However, the following guideline may be used.

(a) 1 - 12" turbine and 6 - under eave soffit vents per 900 square feet attic floor area.

(b) 1 - 14" turbine and 8 - under eave soffit vents per 1250 square feet of attic floor area.

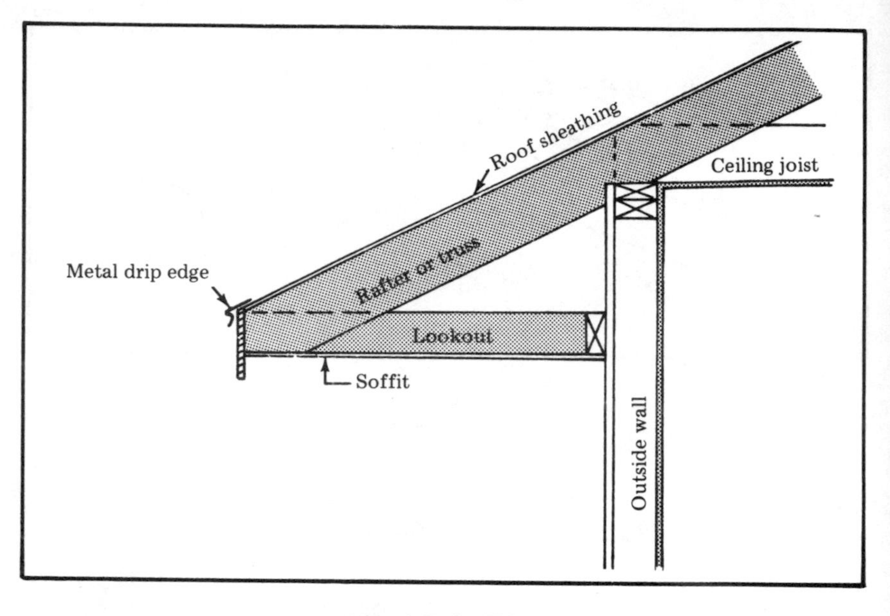

Metal Drip Edge
Figure 8-4

Metal Drip Edge:
The metal drip edge (Figure 8-4), is placed over the roof sheathing and felt along the fascia, and under the roof shingles. It allows water to drip free of the fascia.

When estimating drip edge take the total linear feet of the eaves, divide by 10 (10'0'' lengths) and round off to the next highest whole number for the number of pieces required. *Example:*

Total length of eaves: 128 lineal feet

$$\frac{128}{10} = 12.8 \text{ or } 13 \text{ pieces}$$

Asphalt and Fiberglass Shingles:
About 85% of the houses built in recent years have asphalt shingles for the roof covering. Asphalt shingles are made of organic material, usually wood chips and paper, and impregnated with asphalt and mineral granules, (Figure 8-5). The most commonly used asphalt shingle is a strip shingle 12'' wide and 36'' long. The most common weight is the 235-240 lbs. per square. A 5'' weather exposure requires 80 of these shingles to cover 100 square feet, and they are packed 3 bundles per square. One square of shingles is always the number of shingles required to cover 100 square feet of roof area. *Example:* A strip shingle 12'' x 36'' with a 5'' weather exposure requires 80 shingles to cover 100 square feet.

5'' (exposure) x 36'' (length) x 80 (number of shingles)

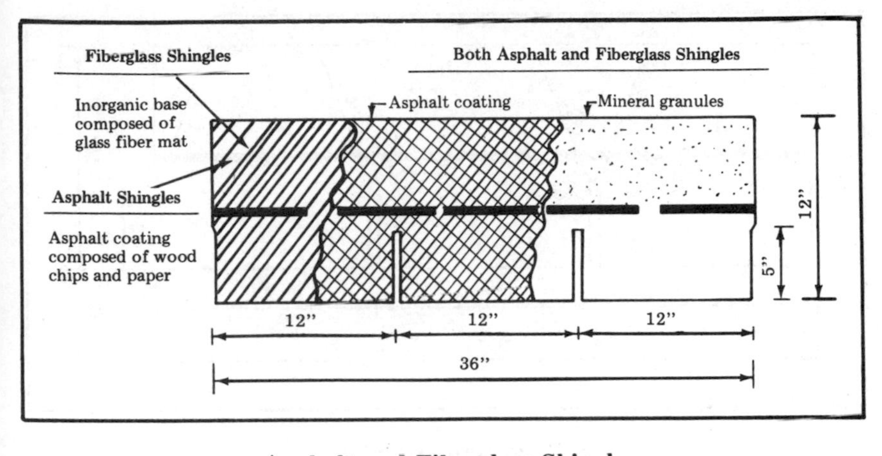

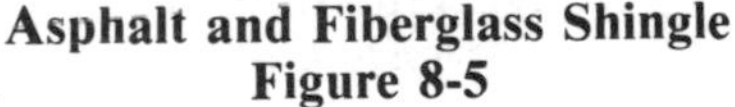

Asphalt and Fiberglass Shingle
Figure 8-5

5" x 36" x 80 = 14,400 square feet

$$\frac{14{,}400}{144\,*} = 100 \text{ square inches}$$

*Note: 1 square foot = 144 square inches

In recent years the fiberglass shingle has accounted for more and more of the asphalt roofing market. In 1975 it accounted for less than 4%. It is estimated that by the mid 1980's it will account for 50% of the market. This shingle looks very much like the asphalt shingle but it has an inorganic base composed of glass fiber mat (Figure 8-5). It is lighter than the asphalt shingle and easier to handle.

To estimate asphalt or fiberglass shingles, calculate the roof area the way roof sheathing was estimated in Chapter 7. Use either the rafter length multiplied by the eave length, or the ratios in Figure 7-24, multiplied by the number of square feet of floor area to determine the roof area. Add an additional allowance to the roof area for the extra course of shingles for the starter strip, and the ridge and the hip caps. Divide the total roof area by 100 for the number of roofing squares required.

Some estimators, in computing the total roof area allow one extra foot for the starter course and one extra foot for each ridge and hip. However, the following factors may be used for estimating the extra shingles required for the starter course using 12" x 36" shingles.

(Factor) .33333 x Total eave length = number of shingles
Example: .33333 x 128 lin. ft. = 42.67 or *43 shingles*
or

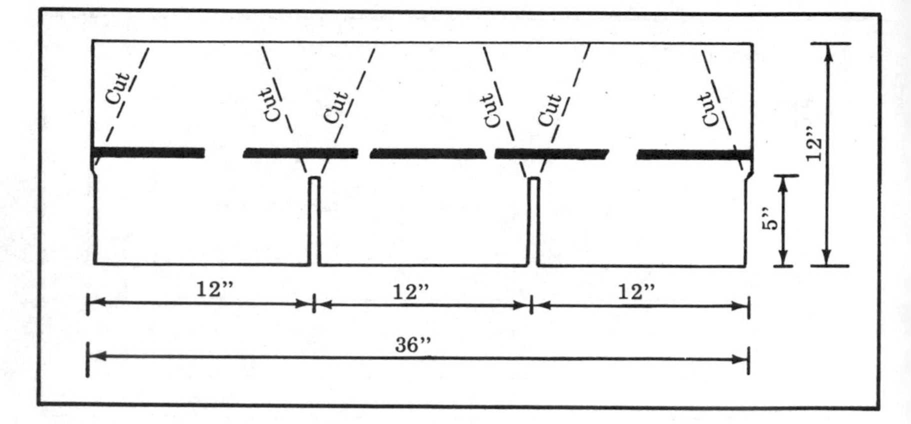

Ridge and Hip Caps
Figure 8-6

(Factor) *.00417 x Total eave length* = *number of squares*
Example: .00417 x 128 lin. ft. = *.53 squares*

Proof: $\frac{42.67 \text{ shingles}}{80*}$ = *.53 squares*

*Note: 80 shingles per square

The following factors may be used for estimating the extra shingles for the ridge and hips using 12" x 36" shingles with a 5" exposure (Note: one shingle makes three caps, see Figure 8-6).

Number of Shingles for Ridge and Hips
(Factor) .80000 x Lineal feet of ridge and hips = number of shingles
Example: .80000 x 52 lin. ft. = 41.6 or 42 shingles

or

Number of Squares for Ridge and Hips
(Factor) .01000 x Lineal feet of ridge and hips = number of squares
Example: .01000 x 52 lin. ft. = .52 squares

Proof: 52 lin. ft. x 12 = 624"

$\frac{624"}{5*}$ = 124.8 pieces or number of caps

*Note: 5" weather exposure

$\frac{124.8}{3**}$ = 41.6 shingles

**Note: One shingle makes three caps

$\frac{41.6}{80***}$ = *.52 squares*

***Note: 80 shingles per square

The number of squares of asphalt roofing required for the house and porch roofs in Figures 7-1, 7-2, 7-10, 7-35 and 7-36 will be:

Roof area of house (From chapter 7)	1984 sq. ft.
Roof area of porches and walks	672 sq. ft.
Total	2656 sq. ft.

$$\frac{2656}{100} = \textit{26.56 squares}$$

Starter Courses

Total length of eaves: 193 lineal feet
(Factor) .00417 x 193 lin. ft. = .80 squares

Ridge

Total length of ridges: 67 lineal feet
(Factor) .01000 x 67 lin. ft. = .67 squares

The total number of squares of asphalt roofing will be:

Main house	26.56 squares
Starter courses	.80 squares
Ridges	.67 squares
	28.03 squares
Add 3% to waste	.84
Total	28.87 squares

Order: 29 - squares 235 lb. asphalt roofing

Nails

Allow:
(a) 1½ lbs. 1" roofing nails, or 1¾ lbs. 1¼" roofing nails per square.

(b) For the ridge and hip caps using 1½" roofing nails multiply 0.02667 (Factor) by total linear feet of ridge and hips for the number of pounds of nails.

(c) For the ridge and hips using 1¾" roofing nails multiply 0.03077 (Factor) by the total linear feet of ridge and hips for the number of pounds of nails.

The nail requirement for 29 squares of asphalt roofing using 1" roofing nails will be:

29 squares x 1½ = 43.50 or 44 lbs.

The nails required for the ridges using 1½" roofing nails will be:

.02667 (Factor) x 67 lin. ft. = 1.79 or 2 lbs.

Order: 44 lbs. 1" roofing nails
2 lbs. 1½" roofing nails

Shingle Length	Minimum Thickness	Approximate Coverage of One Square (4 bundles) Shingles Based on Following Weather Exposure 4"	4½"	5"	5½"	6"	6½"	7"	7½"
16"	5 in 2"	80	90	100*	110	120	130	140	150
18"	5 in 2¼"	72½	81½	90½	100*	109	118	127	136
24"	4 in 2"	---	---	---	---	80	86½	93	100*

Wood Shingle Exposure Chart
Figure 8-7

Wood Shingles:
Prior to 1940 most houses were roofed with wood shingles. Today, however, asphalt shingles are used on approximately 85% of the houses constructed. Wood shingles are sawed from many species of wood, but basic grades of red cedar are: Number 1, Blue Label; Number 2, Red Label; and Number 3, Black Label.

The Number 1 grade is the premium grade, Number 2 is a good grade for most applications and Number 3 is a utility or economy grade.

Each grade comes in lengths of 16", 18" and 24", and in random widths. They are packed 4 bundles per square. The approximate weight per square for each grade is:

Length	Weight
16"	144 lbs.
18"	158 lbs.
24"	192 lbs.

Figure 8-7 is a wood shingle exposure chart showing the approximate coverage of one square (4 bundles) based on the weather exposure from 4' to 7½". The maximum recommended weather exposure for a 16" shingle is 5"; for an 18" shingle, 5½"; and for a 24" shingle, 7½".

When estimating wood shingles, divide the roof area by the coverage for the weather exposure to be used, then add 10% for gable roofs for the starter strip and ridge. For hip roofs add 15% for starter strip, ridge and hips. *Example*: A gable roof with a roof area of 1684 square feet is to be covered with 16" red cedar shingles with a weather exposure of 4½". The coverage for a 16" shingle with a 4½" exposure in Figure 8-7 is 90 (square feet).

Number of Square

$$\frac{1684 \text{ sq. ft.}}{90 \text{ (coverage)}} = 18.71 \text{ squares}$$

	18.71 squares
Add 10% for gable roof	1.87
Total	20.58 or *20¾ squares*

Handsplit and Resawn	Approximate Square Foot Coverage of One Square of Handsplit Shakes Based on These Weather Exposures 5½"	6½"	7"	7½"	8"	8½"	10"
18" x ½" to ¾"	55*	65	70	75**	80	85	--
18" x ¾" to 1¼"	55*	65	70	75**	80	85	--
24" x ½" to ¾"	--	65	70	75*	80	85	100**
24" x ¾" to 1¼"	---	65	70	75*	80	85	100**
32" x ¾" to 1¼"	--	--	--	--	--	--	100*

Note: *Recommended maximum weather exposure for 3-ply roof construction.
**Recommended maximum weather exposure for 2 ply roof construction.

Handsplit Shakes
Figure 8-8

Number of Bundles

20.58 x 4 = 82.32 or *83 bundles*

Example: A hip roof with a roof area of 1802 square feet will be covered with 24" red cedar shingles with a weather exposure of 7½". The coverage for a 24" shingle with a 7½" exposure in Figure 8-7 is 100 (square feet).

Number of Squares

Roof area / Coverage: 1802 sq. ft. / 100 = 18.02 squares

	18.02	18.02
Add 15% for hip roof		2.70
Total		20.72 or *20¾ squares*

Number of Bundles

20.72 x 4 = 82.88 or *83 bundles*

Handsplit and resawn shakes are split wood shingles with a rough texture and are much heavier. Cedar shakes are available only in Number 1 grade. They are sawn in lengths of 18", 24" and 32", and come in random widths. They are packed 5 bundles per square. They vary in weight from approximately 225 lbs. per square for the 18" x ½" x ¾" shake to double that for the 32" x ¾" x 1¼" shake. Figure 8-8 is a chart showing the approximate coverage of one square of handsplit shakes for weather exposures from 5½" to 10".

Estimate handsplit shakes the same as for wood shingles, except use the chart in Figure 8-8. *Example:* A gable roof with a roof area of 1684 square feet is to be covered with 24'' red cedar handsplit shakes with a weather exposure of 10''. The coverage for a 24'' shake with a 10'' weather exposure in Figure 8-8 is 100 (square feet).

Roof area / Coverage	1684 sq. ft. / 100	= 16.84 squares
	16.84	
Add 10% for gable roof	1.68	
Total	18.52 or *18-3/5 squares*	

Number of Bundles

(Handsplit shakes: 5 bundles per square)

18.52 x 5 = 92.60 or *93 bundles*

Nails

Use rust-resistant 1½'' (4d) or 2'' (6d) nails.

Allow:

(a) 2 lbs. 4d nails per square

(b) 3 lbs. 6d nails per square

Built-Up Roofing:

Built-up roofing is a roof covering that consists of several layers of roof felt with each layer mopped on with hot tar or asphalt and topped with an aggregate such as gravel for protection against the weather.

Most built-up roofs are installed on roofs with a pitch of 0 on 12 (flat roof) to 3 in 12. However, they can be installed on roofs up to a 6 in 12 pitch.

The installation of built-up roofing is specialized work and should be done by professionals. For this reason built-up roofing is normally let on subcontract. The estimator should always have a complete specification sheet from the roofing contractor before letting the contract for the built-up roofing.

The material list for the roofing on the house in Figures 7-1, 7-2, 7-10, 7-35 and 7-36 is:

Felt: 7 rolls 15 lb. felt

Nails for Felt: 14 lbs 1'' roofing nails

Flashing: 3 rolls 14'' x 50' aluminum

Ridge Vents: None

Roof Ventilating Louvers: None

Metal Drip Cap: 17 pieces (10' length)

Asphalt Shingles: 29 squares 235 lbs. roofing shingles

Nails: 44 lbs. 1'' roofing nails

2 lbs. 1½'' roofing nails

Wood Shingles: None

Built-Up Roofing: None

Roof Cement: 5 gals. plastic roof cement

**Enter above material on Line 8.1 Cost Estimate Worksheet*

Labor Costs For Roofing

Labor costs for installing roofing vary with the roof pitch, the type of roofing material used, and the skill of the workment. The man-hours per square for a professional roofer can vary from less than one man-hour per square for asphalt shingles installed on a gable roof with a roof pitch not exceeding 6 in 12, up to 6 or more man-hours per square for the installation of the wood shingles on steep roofs where foot rests are required. On mansard roofs (having two pitches on all four sides from the eave to the ridge) the lower pitch may exceed 20 in 12. The man-hours per square may be much higher on these roofs.

The man-hours per square for carpenters who install roofing only occasionally will probably be:

Asphalt shingles	2 - 6 *
Wood shingles	3 - 6 *

* Note: Add up to 50% for steep roofs where foot rests will be required.

Roofing contractors can normally install roofing more cheaply than carpenters. Using roofing contractors releases the carpenter for other work that would otherwise be delayed. Roofing contractors will contract the work on a per job basis, and will either furnish the material and labor, or the labor only for a fixed amount per square.

A complete specification sheet should be signed by the roofing contractor before letting the contract. A guideline that may be used is:

1. Felt paper: Weight

2. Metal drip edging

3. Roof vents: Type and size

4. Flashing: Type and size and where it is to be used, and how it is to be installed. Will valleys be open with the flashing showing, or will they be closed or woven where the roof shingles overlap?

5. Roof shingles: Type and weight

6. Nails: Size and number per shingle

7. Insurance: If the roofing contractor does not have workers compensation and liability insurance on his employees, this insurance should be provided and the cost added to the roofing costs.

The installation of the asphalt roofing on the house in Figures 7-10, 7-35 and 7-36 was let to a roofing contractor for the labor only. His fee was $18.50 per square. The labor cost for installing this roofing was:

*29 squares @ $18.50 = $536.50 **

*Note: This includes the insurance.
(Enter this labor cost on Line 8.2 Cost Estimate Worksheet)

Cost Estimate Worksheet For Roofing

8.1 Roof Covering:

Felt
_________Rolls (weight_________) @_________ = $_________

Nails for Felt
_________lbs. (size_________) @_________ = _________

Flashing
_________Rolls @_________ = _________

(Type_____________; size_____________)

Ridge Vents
_________lin. ft. @_________ = _________

Connector Plugs
_________Number @_________ = _________

End Plugs
_________Number @_________ = _________

Roof Ventilating Louvers
_________Number (size_________) @_________ = _________

Metal Drip Edge
_________lin. ft. @_________ = _________

Roofing Shingles (or Built-Up Roofing)
_________Squares @_________ = _________

(Type_________; Size_________; Weight_________)

Nails
_____ lbs. (size _____) @ _____ = _________
_____ lbs. (size _____) @ _____ = _________
_____ lbs. (size _____) @ _____ = _________

Roof Cement
_____ Gals. (type _____) @ _____ = _________

Total for line 8.1 $ _________ $_________

Sales tax (_____%) _________

Total cost of material $_________

8.2 Labor costs for roofing $_________
(Include insurance for subcontract labor)

Total cost of roof covering $_________
(Enter on Line 8, Form 100)

Chapter 9

Electrical, Plumbing, Heating and Air Conditioning

The installation of the electrical, plumbing, heating and air conditioning is a specialized trade. A license and a permit to do this work are required by most localities. The work must conform to the local building codes, and one or more inspectors will inspect it for compliance. All of the electrical, plumbing, heating and air conditioning is normally done by subcontractors. The rough-in for this work follows the installation of the roof and precedes the finish walls and ceilings.

Electrical

The electrical system provides the house with the electric power appropriate to its needs.

Since all work will be inspected for code compliance, it will be time well spent to take the blueprints to the local electric power company and get their recommendations for the electrical layout of the house before letting the contract out on bids. The electrical layout shown on the plans may not be adequate for the appliances and equipment that are planned for the house. Here the local electric power company can be very helpful, and may save the owner the additional expense of adding to it later.

The electrical contractor will normally bid on the labor and material for the rough-in electrical work, such as wiring, boxes for outlets, switches and fixtures, the entrance panel and circuit breakers and cover plates for the outlets and switches. His bid normally includes hanging the light fixtures (but not furnishing them) and the hook-up for the appliances, such as the water heater, the oven, the dishwasher, that are shown on the plans and specifications. Any special equipment, including the hook-up for the heating, cooling and ventilating system, the installation of smoke detectors, intercom systems, central vacuum systems, sump pumps, door bells, telephone service line and boxes, television service line and boxes should be included in the bid.

Arrive at an understanding with the electrical contractor as to where his responsibility begins. It normally begins at the meter. If the electric service line to the house is underground there will probably be a charge by the electric power company for this service. The distance the house is located from the available source of power may add to this expense.

A checklist to help the builder or estimator for estimating electrical costs follows:

1. Cost of service line to house.

2. Cost of hook-up for temporary electric service.

3. Labor and material for rough-in. This includes wiring, outlets, boxes and plates, boxes for the fixtures, switches, connectors, the entrance panel and circuit breakers. The subcontractor should provide insurance on his employees.

4. Cost of light fixtures.

5. Installation of light fixtures.

6. Hook-up for appliances.

7. Hook-up for heating, cooling and ventilating equipment.

8. Hook-up and installation of special equipment.

9. Telephone boxes and service to house.

10. Television boxes and service to house (if required).

The cost of the electrical system for the house in Figures 7-1 and 7-2 is:

1. Cost of service to house: none.

2. Cost of hook-up for temporary service: included in bid.

3. Labor and material for rough-in: $3200.00

4. Cost of light fixtures: allow $650.00 *

* Note: This includes smoke detectors and central vacuum system.

5. Installation of light fixtures: included in bid.

6. Hook-up for appliances: included in bid.

7. Hook-up for heating, cooling and ventilating equipment (This is included with the electric heat).

8. Hook-up and installation of special equipment (furnished by owner): The following are included in the bid:

(a) Central vacuum system (installation and hook-up).

(b) Hook-up of sump pump in basement.

(c) Installation of six smoke detectors.

9. Telephone boxes and service to house: by local telephone company.

10. Television boxes and television cable service to house: by local TV cable company.

The total cost of the electrical system is:

Contract	$3200.00
Fixtures and special equipment	650.00
Total	$3850.00

(Enter on Line 9.1 Cost Estimate Worksheet)

Plumbing

The plumbing should be designed to provide adequate water, proper drainage and venting for the dwelling.

As with the electrical system, most areas require a license and permit to do plumbing work. All plumbing must comply with the local building codes and an inspector will inspect the work for compliance. If the house is built where there is no public water service, or a public sewer system, the local health department will probably have the final authority over the private water supply and the private sewerage discharge system that will be required.

The take-off for the rough-in material for the plumbing is work for experienced personnel, and should be done by the plumbing contractor. The fixtures, such as the kitchen sink, garbage disposal, bathtubs, shower stalls, water closets, lavatories and faucets are normally selected by the owner and installed by the plumber.

Before letting the plumbing contract be sure the specifications state the type and size of the water supply pipes and the drain pipes. If the plumber is to furnish any fixtures, including the water heater, the specifications should make clear the type and size fixture to be furnished. The fixtures furnished by the owner should be included in the specifications and the responsibility for installing them made clear. The limits of responsibility for the water supply lines and the sewer lines outside of the house should also be included in the specifications. Some plumbing contractors will bid on the work to a specified distance outside of the foundation 5 or 10 feet. The cost of extending the water supply and the sewer lines to the point of connection and discharge will be extra. The responsibility for digging the water and sewer lines and covering them should be defined in the specifications. Seldom will the plumbing contractor extend his area of responsibility beyond the property line without a separate bid.

A checklist to help estimate plumbing costs follows:

1. Cost of water supply line from point of connection to the house, including the trench for the water pipe.

2. Temporary water service hook-up.

3. Cost of sewer line from house to point of discharge, including the trench for the sewer pipe.

4. Cost of labor and material for rough-in. This includes type and size of water supply pipes, type and size of drain pipes, type and size of vent pipes, number and location of outside water faucets and all material required to make the necessary connections.

5. Cost of plumbing fixtures.

6. Installation of plumbing fixtures.

7. Installation of special equipment such as water softeners, pressure relief valves, sump pumps, shower doors, etc.

The cost of the plumbing for the house in Figures 7-1 and 7-2 is:

1. Cost of water supply line from point of connection to the house, including the trench for the water line: included in bid.

2. Temporary water service hook-up: included in bid.

3. Cost of sewer line from house to point of discharge, including the trench for the sewer pipe: included in bid.

4. Cost of labor and material for rough-in: $2883.00 (This includes the trenches for the water and sewer lines).

5. Cost of plumbing fixtures: Allow $1500.00 for three bathrooms (includes three shower doors), kitchen, utility area, water heater and a sump pump in the basement.

6. Installation of plumbing fixtures: included in bid.

7. Installation of special equipment: The installation of the sump pump in the basement is included in the bid. The shower doors are to be installed by others. There is no other special equipment.

The total cost of the plumbing system is:

Contract	$2883.00
Fixtures and special equipment	1500.00
Total	$4383.00

(Enter on Line 9.2 Cost Estimate Worksheet)

Heating and Air Conditioning

Heating

The heating system in a house should be efficient, safe and convenient to use.

There are many types of heating systems in residential buildings. There are oil, gas, electric and bio-fueled furnaces, space heaters, electric radiant heat with heating cables in the ceiling, heat pumps that are also air conditioners, and in increasing numbers, solar heaters.

Regardless of which type of heating system to be used, it should be designed and installed by a reputable heating contractor to meet all applicable building codes. If the heating unit is too small or not installed properly it will not provide sufficient heating; if the heating unit is too large it will be unnecessarily expensive to purchase and operate.

The size of the heating unit will depend on the size of the house and the calculated heat loss. Adequate insulation in the house and less window area will reduce the heat loss, thus resulting in a savings in the operating costs and a reduction in the size of the heating unit needed. For example, if electric baseboard heat or electric radiant heat is used the size of the heating unit will be computed at approximately 6 to 10 watts per square foot of living area depending on the heat loss. A room with 200 square feet and a high heat loss factor will probably require 2000 watts of heat, whereas the same room with a low heat loss factor would probably be computed between 1200 and 1500 watts of heat. This will result in a savings in both the cost of the heating unit and in the operating costs.

Air Conditioning

The air conditioning facilities should be safe, quiet, economical and effective. The size of the air conditioning system, as with the heating system, will depend on the size of the house and the calculated heat loss, and should meet all applicable building codes.

The size of thru-the-wall and window type air conditioners are normally measured in B.T.U.'s per hour and are used to cool a designated area in the building. Central air conditioning is measured in tons of refrigeration (one ton equals 12,000 B.T.U.'s per hour) and is designed to cool the entire building, or a major part of the building, such as one floor.

In recent years the heat pump has gained wide acceptance in residential construction. It is both a heating and cooling system. It supplies heat in the winter and air conditioning in the summer by reversing the cycle. It has a built-in auxiliary resistance heating unit that automatically switches on when the outside temperature is too cold for the heat pump to operate efficiently. Heat pumps are more efficient to operate than resistant type heaters.

Heat pumps and other central air conditioning should be designed and installed by a reputable heating and air conditioning contractor. For a preliminary estimate the builder or estimator may use as a rule-of-thumb the formula: 1 ton refrigeration (12,000 B.T.U.'s) per 600 square feet, to estimate the size of heat pumps that may be required for a house.

Thus, a house with 1800 square feet of living area on one floor will probably require a 3 ton heat pump, and a two story house with 1200 square feet of living area on the first floor and 800 square feet of living area on the second floor will probably require a 2 ton heat pump for the first floor and a 1½ ton heat pump for the second floor.

The house in Figures 7-1 and 7-2 will use electric base board heaters, and the owner will furnish three thru-the-wall air conditioners. The labor and material for all heat and wiring for the thru-the-wall air conditioners was let on contract for: $1875.00

(Enter this amount on Line 9.3 Cost Estimate Worksheet)

Cost Estimate Worksheet For Electrical, Plumbing, Heating and Air Conditioning

9.1	Electrical	$________
9.2	Plumbing	________
9.3	Heating and air conditioning	________
	Labor for carpenters *	________
	Total	$________

(Enter on Line 9, Form 100)

* Note: Coordination work with the subcontractors such as cutouts, openings, supports, etc.

Chapter 10
Estimating Brickwork

Before the brickwork begins all window and door frames where the brick is used should be installed. The ground where the brick masons will be working should be leveled and all materials necessary for the completion of the brickwork such as brick, masonry cement, sand, water, brick lintels, wall ties, anti-freeze (if needed), and scaffold material should either be on the job site, or ordered, with the assurance they will be delivered before the brick masons need them.

The type and description of the brick is the decision of the home owner, and the price of the brick per M (thousand) depends on the brick type and description. Often the owner has not decided on the brick at the time the estimate is prepared. When this happens an allowance must be made on the cost of the brick per M for each different type of brick the owner may be interested in. The Standard brick is the most commonly used brick in residential construction. The size of this brick is 2¼'' high (face of brick) x 3¾'' wide (bed) x 8'' long. Other bricks that are popular in residential construction are Norman bricks (2¼'' x 3¾'' x 11½''), and Roman bricks, which have a long lean dimension (1½'' x 3¾'' x 11½''). Another brick that was originally intended for one-story houses but is more commonly used in one-story commercial buildings is the SCR brick. This brick is 2¼'' x 5½'' x 11½'' and builds a single wall with a nominal dimension of 6 inches wide. These bricks do not require any frame or masonry back-up, but are merely furred for drywall or plaster. Their advantage is that only 4.40 units are required per square foot, thus increasing production. Their disadvantage is the high heat loss through the walls.

Bricks may vary slightly in size due to variations in the clay and the firing temperature. Overburned bricks will be smaller than underburned bricks. The color may also vary. An average difference of ¼ inch can increase the number of bricks required. The estimator or builder should keep this in mind when ordering brick.

Figure 10-1 is a chart showing the brick factor per square foot, the masonry cement factor per brick and the sand factor per brick for Standard bricks, Roman bricks, Norman bricks and SCR bricks. This chart may be used in estimating the number of bricks, bags of masonry cement and tons of sand. An example of the computations for the brick factors in Figure 10-1 follows.

The masonry cement factors are computed on .50 bags per cubic foot of mortar, which allows for waste. *No additional allowance should be made for masonry cement waste.*

The sand factors are computed for dry sand with a weight of 100 lbs. per cubic foot of mortar. *For wet sand an additional allowance of 25% to 40% should be made in the weight of the sand. Example:*

6550 bricks x .00047 = 3.08 tons sand
Add 25% for waste = .77 tons sand
3.85 (order 4 tons sand)

Explanation

1. *Number of bricks:* (equals) Square feet of wall area multiplied by brick factor in either column [a] or [b].

2. *Number bags masonry cement:* (equals) Number of bricks multiplied by masonry cement factor in either column [c] or [d].

3. *Tons of sand:* (equals) Number of bricks multiplied by sand factor in either column [e] or [f].

Example

1000 square feet of wall area using Standard Brick w/⅜" joints.

1. 1000 x 6.55 (Brick factor for ⅜" joint in column [a]) = 6550 Bricks.

2. 6550 Bricks x .00466 (Masonry cement factor for ⅜" joint in column [c]) = 30.52 Bags Masonry Cement.

3. 6550 Bricks x .00047 (Sand factor for ⅜" joint in column[e]) = 3.08 Tons Sand.

Example

6380 square feet wall area using Norman Brick w/½" joints

1. 6380 square feet x 4.40 (Brick factor for ½" joints in column [b]) = 28,072 Bricks

2. 28,072 Bricks x .00818 (Masonry Cement factor for ½" joints in column [d]) = 229.63 Bags Masonry Cement

3. 28,072 Bricks x .00082 (Sand factor for ½" joints in column [f]) = 23.02 Tons Sand

Brick and Mortar Factors For Single Wall

Type of Brick	Brick Factor Per Square Foot		Masonry Cement Factor (Bags) Per Brick		Sand Factor (Tons) Per Brick	
	3/8" Joint (a)	1/2" Joint (b)	3/8" Joint (c)	1/2" Joint (d)	3/8" Joint (e)	1/2" Joint (f)
Standard Brick 2¼" x 3¾" x 8"	6.55	6.16	.00466	.00625	.00047	.00063
Roman Brick 1½" x 3¾" x 11½"	6.40	6.00	.00531	.00750	.00053	.00075
Norman Brick 2¼" x 3¾" x 11½"	4.60	4.40	.00587	.00818	.00059	.00082
SCR Brick 2¼" x 5½" x 11½"	--	4.40	--	.01148	--	.00115

Note: This brick factor *does not* include an allowance for waste. An additional allowance of of 3% to 5% should be made for brick waste. *Example:*

1000 square feet x 6.55 (Factor) = 6550.00 bricks
Add 5% for waste = 327.50 bricks
6877.50 (order 6900 bricks)

Brick and Mortar Factors for Single Wall
Figure 10-1

The brick factor for Standard brick for a ½" joint in column [b] is 6.16 bricks per square foot. The factor was computed as follows:

(a) The Standard brick is 2¼" plus one ½" joint equals 2¾" in height; the brick length is 8" plus one ½" joint is 8½".

$$2.75" (2\tfrac{3}{4}") \times 8.5" (8\tfrac{1}{2}") = 23.38 \text{ sq. in.}$$

$$\frac{23.38 \text{ sq. in.}}{144^*} = .16236 \text{ sq. ft.}$$

***Note:** 1 square foot = 144 square inches

Solution (using proportion)

1 brick: .16236 sq. ft. = x bricks: 1 sq. ft.

$$\overbrace{1 : x}^{\text{Bricks}} = \overbrace{.16236 : 1}^{\text{Square Feet}}$$

x = *6.16 bricks per square foot*

The other brick factors were computed in a similar manner for the size of each brick type and the size of the mortar joints shown.

The masonry cement factors were computed using .50 bags per cubic foot of mortar, allowing for waste (.36 bags per cubic foot is the actual count but does not allow for waste).

The weight of sand can vary from approximately 90 lbs. per cubic foot for dry sand to approximately 140 lbs. per cubic foot for wet sand. The sand factor in Figure 10-1 is computed for sand weighing 100 lbs. per cubic foot. An allowance of 25% to 40% should be added for wet sand.

Figure 10-2 is a chart showing factors that may be used in estimating the number of brick courses for any wall height using Standard, Norman, SCR and Roman bricks. Figure 10-3 is an illustration for two calculations in Figure 10-2. Figure 10-5 is a conversion chart designed to be used for the mathematics using the factors in Figure 10-2.

Brick veneer may cover all or a portion of the exterior walls on residential construction. Check the elevation sections of the blueprints and the specifications for this information. Estimate the brickwork as accurately as possible. Estimating too high may result in the loss of the job; estimating too low may result in a financial loss for the builder. One item that may be overlooked when estimating the brick costs is the tooling of the mortar joints. Most home owners never give a thought about how they want the mortar joints tooled until after the brickwork has been started, and then it may be too late. Figure 10-4 shows the most commonly tooled mortar joints. Get the owner to decide how he wants the mortar joints tooled before the brickwork begins, and pass this information to the brick masons.

Type of Brick	Height of Wall Multiplied by the Following Factors Equals Number of Brick Courses 3/8" Joint	1/2" Joint
Standard brick 2¼" x 3¾" x 8"	4.57143	4.36364
Norman brick 2¼" x 3¾" x 11½"	4.57143	4.36364
SCR brick 2¼" x 5½" x 11½"	4.57143	4.36364
Roman brick 1½" x 3¾" x 11½"	6.40000	6.00000

Example

Norman Bricks With ½" Joints

Wall Height *	x	Factor	=	Number of Brick Courses
0' 2¾" (.229') *	x	4.36364	=	1 course
0' 11" (.917') *	x	4.36364	=	4 courses
7' 4" (7.333') *	x	4.36364	=	32 courses
11' 0" (11.0')	x	4.36364	=	48 courses

Example

Standard Brick With ½" Joints

0' 2¾" (.229') *	x	4.36364	=	1.00 course
1' 0" (1.0')	x	4.36364	=	4.36 courses **
3'-0" (3.0')	x	4.36364	=	13.09 courses **
7'-0" (7.0')	x	4.36364	=	30.55 courses
14' 0" (14.0')	x	4.36364	=	61.09 courses

Example

Roman Brick With 3/8" Joints

0' 1-7/8" (.156') *	x	6.40000	=	1 course
3' 9" (3.75') *	x	6.40000	=	24 courses
7' 6" (7.5') *	x	6.40000	=	48 courses
8' 0" (8.0')	x	6.40000	=	51.20 courses
11' 3" (11.25') *	x	6.40000	=	72 courses

* Note: See conversion chart in Figure 10-5.
** Note: Illustration shown in Figure 10-3.

Factors for Estimating Brick Courses
Figure 10-2

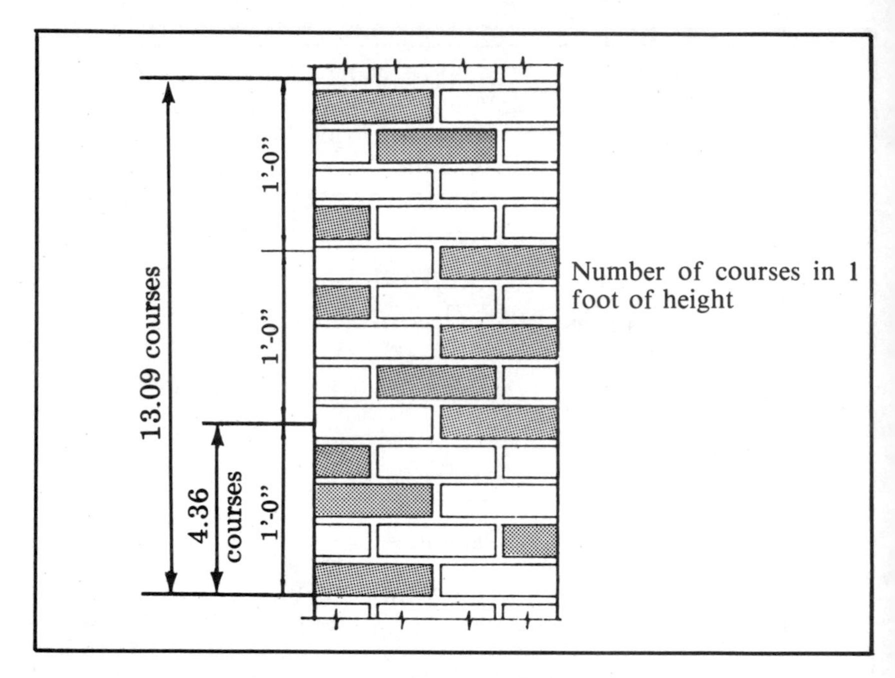

Standard Brick with ½" Joints
Figure 10-3

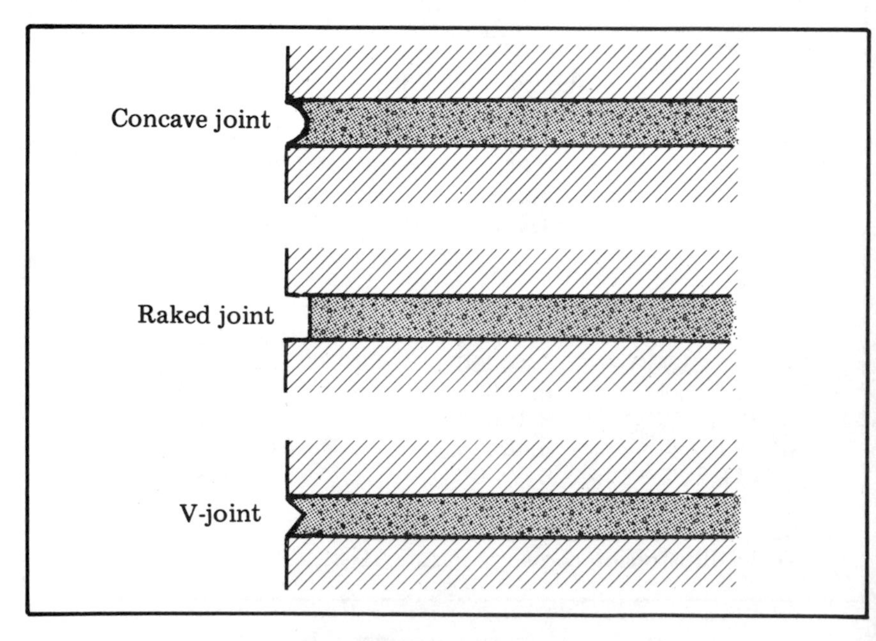

Mortar Joints
Figure 10-4

4th	8th	16th	0"	1"	2"	3"	4"	5"	6"	7"	8"	9"	10"	11"
		0	.000	.083	.167	.250	.333	.417	.500	.583	.667	.750	.833	.917
		1	.005	.089	.172	.255	.339	.422	.505	.589	.672	.755	.839	.922
	1	2	.010	.094	.177	.260	.344	.427	.510	.594	.677	.760	.844	.927
		3	.016	.099	.182	.266	.349	.432	.516	.599	.682	.766	.849	.932
1	2	4	.021	.104	.188	.271	.354	.438	.521	.604	.688	.771	.854	.938
		5	.026	.109	.193	.276	.359	.443	.526	.609	.693	.776	.859	.943
	3	6	.031	.115	.198	.281	.365	.448	.531	.615	.698	.781	.865	.948
		7	.036	.118	.203	.286	.370	.453	.536	.620	.703	.786	.870	.953
2	4	8	.042	.125	.208	.292	.375	.458	.542	.625	.708	.792	.875	.958
		9	.047	.130	.213	.297	.380	.464	.547	.630	.714	.797	.880	.964
	5	10	.052	.135	.219	.302	.386	.469	.552	.635	.719	.802	.885	.969
		11	.057	.141	.224	.307	.391	.474	.557	.641	.724	.807	.891	.974
3	6	12	.063	.146	.229	.313	.396	.479	.563	.646	.729	.813	.896	.979
		13	.068	.151	.234	.318	.401	.484	.568	.651	.734	.818	.901	.984
	7	14	.073	.156	.240	.323	.406	.490	.573	.656	.740	.823	.906	.989
		15	.078	.161	.245	.328	.411	.495	.578	.661	.745	.828	.911	.995

Example: 8¼" = .688'

Decimal Equivalents of Fractional Parts of a Foot
Figure 10-5

Fireplaces and Chimneys

The masonry fireplace (Figure 10-6) that starts at the footer and terminates with the chimney cap above the roof, has practically been replaced by the prefab fireplace (Figure 10-8) in residential construction. The prefab fireplace is energy efficient, easy to install, and can be finished with the same mantle, hearth, and chimney top above the roof as the masonry fireplace, at a fraction of the cost. Regardless of which type of fireplace that is used, the chimney should extend at least two feet above the roof ridge or raised part of the roof, within ten feet of the chimney (Figure 10-9). Figure 10-7 shows different sizes of chimneys and the number of bricks required for each size per foot in height.

The material for the brickwork for the house in Figures 7-1 and 7-2 was estimated as follows:

Standard Brick With ½" Joints are used

1. The brick veneer extended from an average of 2'0" below grade up to the second floor around the perimeter of the house and garage. The perimeter of the house is 192'0" and the height of the walls with brick veneer is 11'0".

192.0' (192'-0" x 11.0' (11'-0")	=	2112.00	square feet
* Less two garage door openings	=	(126.00)	square feet
* Less picture window opening	=	(55.00)	square feet
* Less main entrance door opening	=	(42.00)	square feet
Net	=	1889.00	square feet

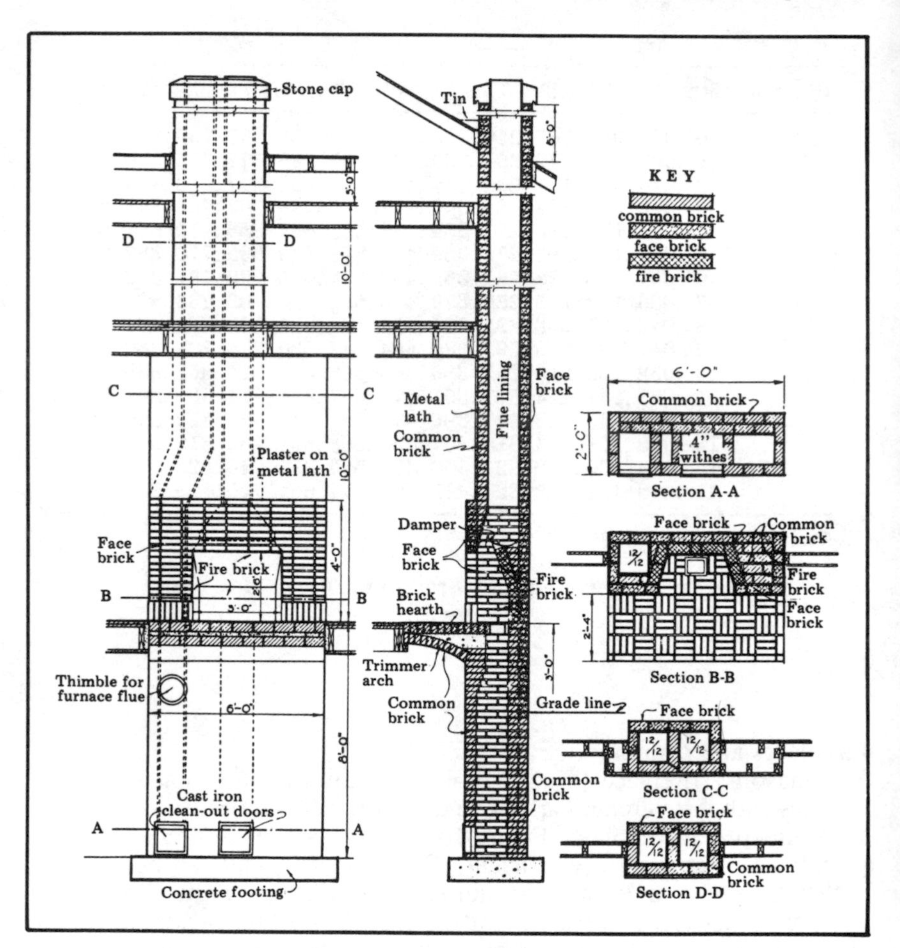

Fireplace Elevation
Figure 10-6

1889 square feet x 6.16 factor (From Figure 10-1)
1889 x 6.16 = *11,636.24 bricks*

2. A double brick retaining wall and brick steps will be built leading from the basement door to the back yard. The retaining wall is 14'6'' in length and 7'0'' high, and there are 10 steps 3'0'' wide.

14.5' (14'-6") x 7.0' x 2 (double wall)	=	203.00 sq. ft.
203 sq. ft. x 6.16 factor	=	1250.48 bricks
10 steps x 58 bricks per step	=	580.00 bricks
Total for wall and steps	=	1830.48 bricks

3. The areaways for three basement windows are to be constructed of brick. The total length of each areaway is 5'0'' and the height is 2'0''.

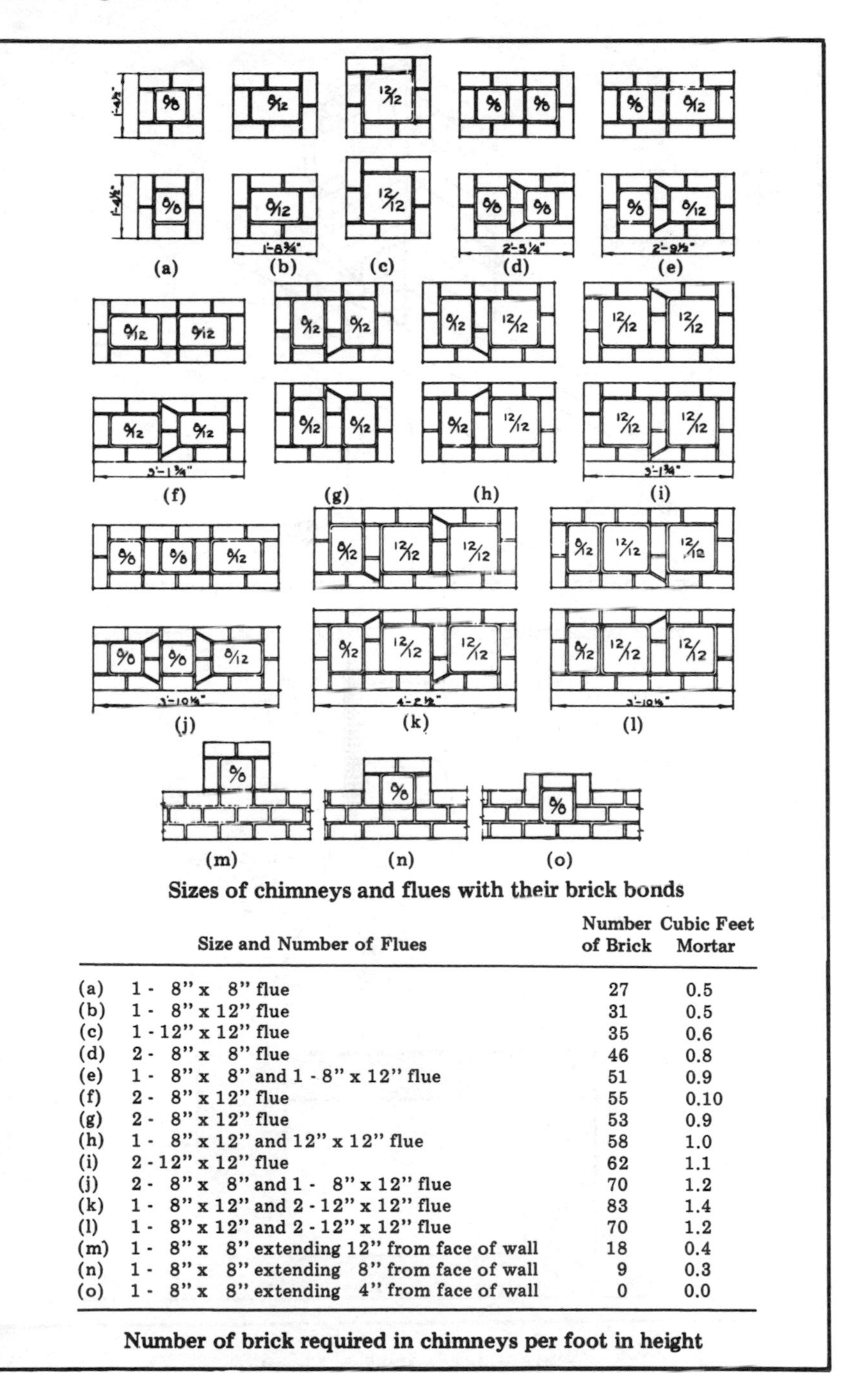

Sizes of chimneys and flues with their brick bonds

	Size and Number of Flues	Number of Brick	Cubic Feet Mortar
(a)	1 - 8" x 8" flue	27	0.5
(b)	1 - 8" x 12" flue	31	0.5
(c)	1 - 12" x 12" flue	35	0.6
(d)	2 - 8" x 8" flue	46	0.8
(e)	1 - 8" x 8" and 1 - 8" x 12" flue	51	0.9
(f)	2 - 8" x 12" flue	55	0.10
(g)	2 - 8" x 12" flue	53	0.9
(h)	1 - 8" x 12" and 12" x 12" flue	58	1.0
(i)	2 - 12" x 12" flue	62	1.1
(j)	2 - 8" x 8" and 1 - 8" x 12" flue	70	1.2
(k)	1 - 8" x 12" and 2 - 12" x 12" flue	83	1.4
(l)	1 - 8" x 12" and 2 - 12" x 12" flue	70	1.2
(m)	1 - 8" x 8" extending 12" from face of wall	18	0.4
(n)	1 - 8" x 8" extending 8" from face of wall	9	0.3
(o)	1 - 8" x 8" extending 4" from face of wall	0	0.0

Number of brick required in chimneys per foot in height

Chimney Sizes and Brick Requirements
Figure 10-7

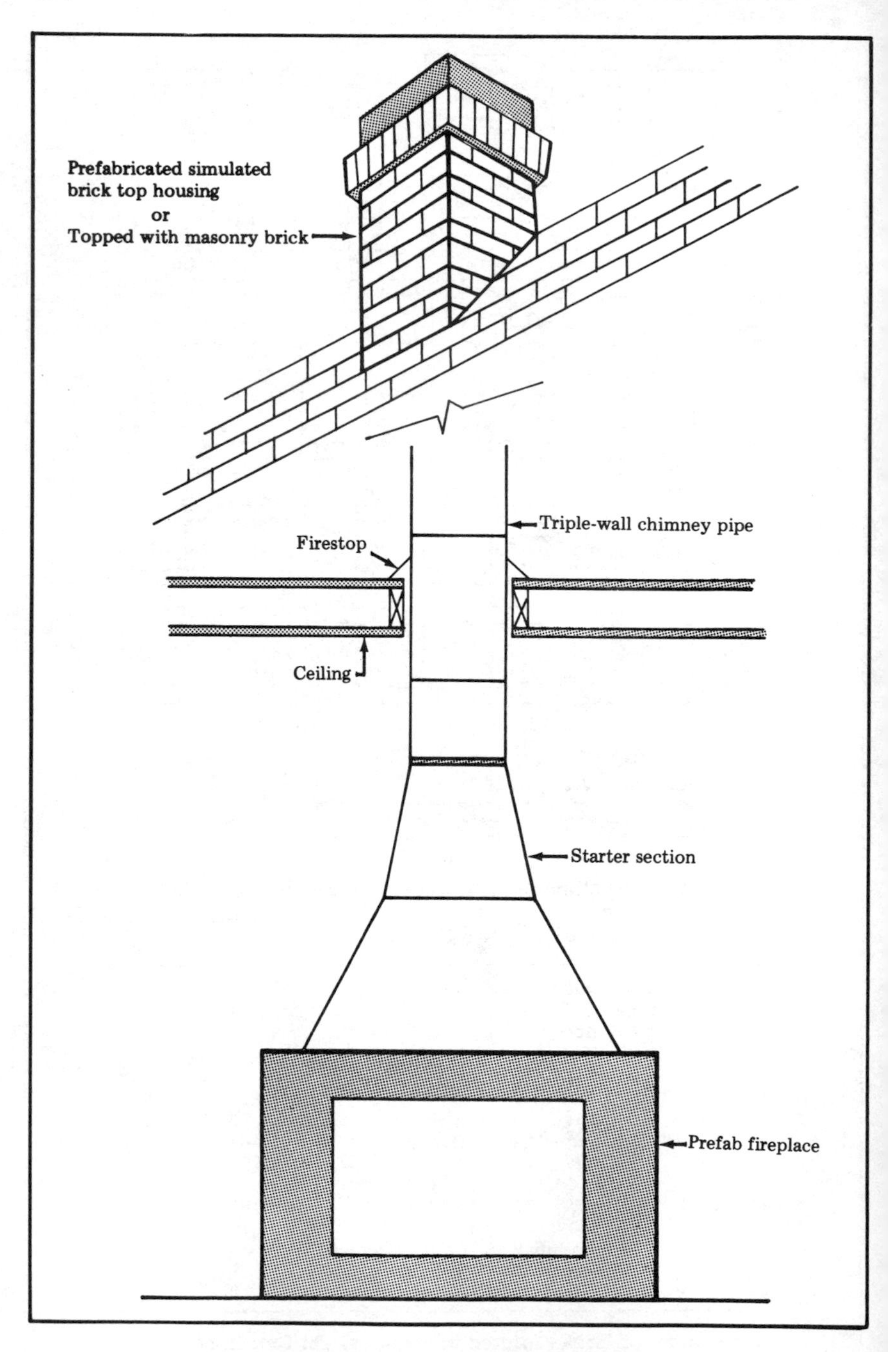

Chimney for Prefab Fireplace
Figure 10-8

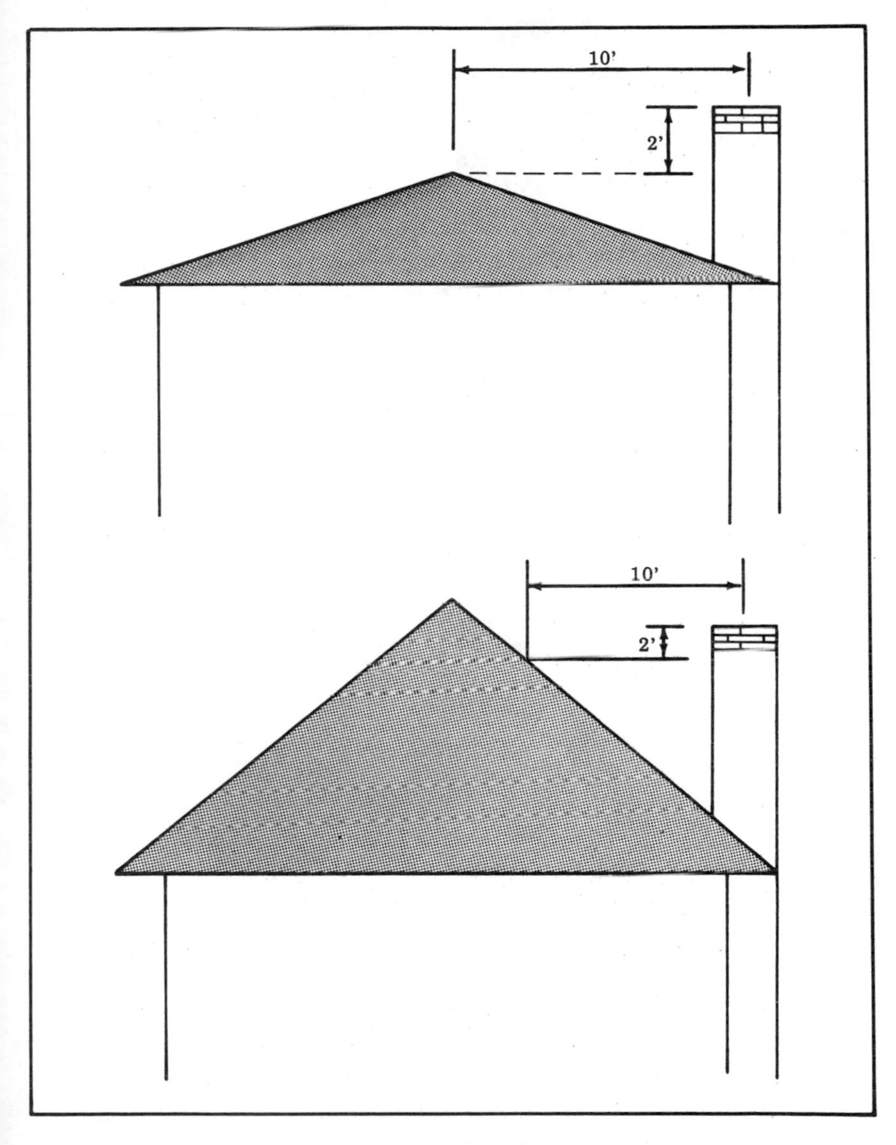

Chimney Heights
Figure 10-9

5.0' (5'0") x 2.0' (2'0") x 3 (areaways) = 30.00 sq. ft.
30 sq. ft. x 6.16 factor = 184.80 bricks

4. The chimney for the prefab fireplace is topped with brick. They extend from below the roof to 2'0" above the ridge for a total height of 5'0". From Figure 10-7 (line c), 35 bricks per foot in height is used for the brick factor.

5.0'' x 35 factor = 175.00 bricks

5. Brick was used on the wall surrounding the prefab fireplace and for the raised hearth. The wall dimension for the brick is 6'0'' x 8'0''

6.0' (6'0'') x 8.0' (8'0'') = 48 square feet
48 square feet x 6.16 factor = 295.68 bricks

The hearth dimension is 6'-0'' x 16'' (1'-4''). It will be a raised hearth (2 courses plus 1 course on edge). 6.0' x 1.33' (16'') = 8.00 square feet
Note: Estimate 4 bricks per square foot for standard brick laid flat, and 6.34 bricks per square foot for brick laid on edge.

8 square feet x 4 bricks x 2 courses	= 64.00 bricks
8 square feet x 6.34 bricks x 1 course	= 50.72 bricks
Total for hearth	=114.72 bricks

The total bricks required for the house are:

1. Mainhouse	11,636.24
2. Retaining wall and steps	1,830.48
3. Areaways	184.80
4. Chimney	175.00
5. Fireplace wall	295.68
6. Hearth	114.72
	14,236.92
Add 5% for waste	711.85
	14,948.77

Order: 15,000 bricks

Prefab Fireplace, including all chimney material: (Order 1)
Masonry Cement:
15,000 bricks x .00625 factor (column [d], Figure 10-1)
15,000 x .00625 = 93.75 bags

Order: 94 bags masonry cement

Sand:
15,000 bricks x .00063 factor (column (f), Figure 10-1)

15,000 x .00063 =	9.45 tons
Add 25% for waste (wet sand) =	2.36 tons
	11.81 tons

Order: 12 tons sand

Brick Lintels:
Order
2 - 10'0''
1 - 8'0''
3 - 6'6''
1 - 5'0''

9 - 4'0"
4 - 3'6"
Note: No brick lintel is required over picture window.

Wall Ties:
Order: 2 boxes (500 each)

Flue Liners:
Order: 1 - 12" x 12" (used for chimney top)

Cleaning Material:
Order: 6 gals. muriatic acid
2 - 6" brushes

Note: No anti-freeze is required for the mortar.

(Enter the above material on Line 10.1 Cost Estimate Worksheet)

Labor Costs For Brickwork

The labor for the brickwork is normally let on subcontract to masonry contractors. However, some builders pay the brick masons and their helpers by the hour. If the masons work by the hour, the builder will have to pay F.I.C.A. and F.U.T.A. taxes, plus Worker's Compensation and Liability Insurance on them. This cost must be included in the labor estimate for the brickwork.

For subcontract labor the masonry contractor will furnish the brick masons and their helpers. He will furnish the scaffolding, mortar mixer, mortar boxes, mortar boards and all tools and equipment necessary for their work. The masonry contractor will either charge a fixed fee per M (thousand) brick, plus an extra charge for such work as the fireplace, hearth, and brick steps, or he will contract the complete job per plans and specifications for one price. The estimator or builder should have an understanding with the masonry contractor on how the mortar joints will be tooled, and who will clean the brick when the brickwork is finished. It is also wise to get a copy of his insurance for verification.

Labor output on brickwork will vary depending on the efficiency of the workmen, weather conditions and the size and shape of the building. The number of corners and window and door openings, the height of the wall, the type of brick, the type of bond and the job conditions are contributing factors to the number of brick a mason can lay in one eight hour day.

Figure 10-10 is a chart that may be used in estimating the man-hours per 1000 brick for Standard, Roman, Norman and SCR bricks. The most accurate labor estimate any estimator can use is always cost records from his previous jobs, but in the absence of these records, this chart may be used.

In addition to the masonry labor there are other labor costs for coordination work with the brick masons that must not be overlooked. For

Type of Brick	Man-Hour Factor Per 1000 (M) Bricks (Mason and Helper)
Standard brick	26.667
Roman brick	50.000
Norman brick	38.636
SCR brick	34.091

Formula: Number of bricks multiplied by factor equals total man-hours.

Example: Estimate the masonry cost to lay 8500 standard bricks.

$$\frac{8500 \text{ bricks}}{1000} = 8.50 \text{ M bricks x } 26.667 \text{ factor} =$$

226.67 or *227 man-hours*

If one brick mason and one helper are used the man-hours will be estimated as follows:

$$\frac{227 \text{ Man-hours}}{2 \text{ workmen}} = 113.50 \text{ man-hours each.}$$

Pay scale for brick mason is $12.50 per hour
Pay scale for helper is $7.50 per hour

113.50 hours @ $12.50	=	$1418.75
113.50 hours @ 7.50	=	851.25
		$2270.00 *

**Note:* Add the cost of payroll taxes and insurance to this amount if the brick mason does not have insurance.

Estimating Man-Hours per 1000 Brick
Figure 10-10

example, the carpenter will have the responsibility for the openings for the chimney, or for the framing around the prefab fireplace. The brick will probably have to be cleaned by the carpenter helpers.

The masonry labor costs for the house in Figures 7-1 and 7-2 were let on contract to a masonry contractor for:

1. Main house and areaways	$2850.00
2. Fireplace wall, hearth and chimney	450.00
3. Retaining wall and brick steps	400.00
Total for masonry labor	$3700.00 *

(Enter this amount on Line 10.2 Cost Estimate Worksheet)

The other labor costs are estimated at.............................$320.00

(Enter this amount on Line 10.2 Cost Estimate Worksheet)

**Note:* The masonry contractor furnished a copy of his insurance with his bid. This price does not include cleaning the bricks. This will be done by the carpenter helpers and this estimated labor cost is included in the other labor costs listed above.

Cost Estimate Worksheet For Brickwork

10.1 Estimating Brickwork:

Brick

________ @ ________ M = $________

(Type: ____________________)

Fireplace (Enter cost of all material below)

Fireplace and chimney material $________

(Type:________) ________

Hearth (Type ________) ________

Mantle (Type ________) ________

$________ ________

Masonry Cement

________ Bags @________ = ________

Sand

________ Tons @ ________ = ________

Brick Lintels

________ (Size ________) @ ________ = ________

________ (Size ________) @ ________ = ________

________ (Size ________) @ ________ = ________

________ (Size ________) @ ________ = ________

________ (Size ________) @ ________ = ________

________ (Size ________) @ ________ = ________

________ (Size ________) @ ________ = ________

________ (Size ________) @ ________ = ________

________ (Size ________) @ ________ = ________

________ (Size ________) @ ________ = ________

Wall Ties
_________ Boxes/M @ _________ = _________

Flue Liners
_________ (Size _________) @ _________ = _________
_________ (Size _________) @ _________ = _________

Cleaning Material
_________ Gals. Muriatic Acid @ _________ = _________
_________ Brushes @ _________ = _________
_________ Other Material @_________ = _________

Mortar Antifreeze
_________ Gals. @ _________ = _________

$ _________ $ _________
Sales tax (____%) _________

(Line 10.1) Total cost of material $ _________

10.2 Labor Costs For Brickwork:
Masonry $_________
Other Labor * _________

(Line 10.2) Total cost of labor $_________ $ _________

Total cost of brickwork $ _________
(Add Lines 10.1 and 10.2)
(Enter on Line 10, Form 100)

* **Note:** Coordination work with brick masons such as cut-outs for chimney openings, framing around fireplace, etc. Cleaning brick will also be included here if it is not done by the masonry contractor.

Chapter 11

Energy Saving Materials

The cost of energy for heating and cooling is going to increase far more rapidly than it has ever done before. If we continue to build homes with the same low insulation standards as before the pre-energy crisis, and energy costs continue to rise as predicted, few people will be able to afford their home heating and cooling costs by the year 2000.

Heat flows to colder objects regardless if they are up, down or through the walls, and this heat loss is measured in Btu's (British Thermal Units) per hour. It may be written Btu/hr or BTUH. One Btu is the amount of heat required to raise one pound of water one degree Fahrenheit. Thus, if a wall has a heat loss of 1500 Btu's, it is losing heat in an amount which would be equivalent to the heat required to heat 1500 lbs. of water 1 °F. The heating system must replace this heat loss to maintain the required temperature within this space. *Therefore, this heat loss must be kept to a minimum for efficient fuel conservation.*

The most important single step in residential energy conservation is proper insulation. Insulation resists the flow of heat, either out of the home in the winter or into the home in the summer, thus significantly reducing the amount of energy consumed in heating and cooling. The ability of an insulating material to resist this heat transfer is known as its resistance value, or R-value. *Industry standard R-values are numbered; the higher the number, the more effective the insulation.* An investment in the cost of insulation and other energy saving materials pays dividends as long as the house stands, in reduced utility bills. In time these dividends pay off the investment. At current fuel prices the energy saving materials will pay for themselves many times over during the life of the house.

The amount of energy that can be saved with the proper use of insulation, storm sash and storm doors can best be illustrated by a house built before the pre-energy crisis. This house, at the time it was built, had full thick insulation in the ceiling only. No insulation was installed in the walls or floors, and no storm sash (the windows were single pane) or storm doors were used. After two years, four additional inches of insulation were blown into the attic, the walls were insulated with blown insulation and the floors (over an unheated basement) were insulated with full thick batts. Storm sash and storm doors were installed on all windows and doors. A record was kept for an eleven year period afterwards (the same family lived in this house throughout this period) and the fuel consumption was reduced by an average of 46.19% per year. This fuel savings, plus the additional comfort of the house, is a dividend on the investment that will continue as long as the house stands.

Types of Insulation

There are many different types of insulation. (See Figure 11-10 for their R-values and thickness.) Some of the most common are discussed below.

Glass Fiber and Rock Wool is available in batts, blankets and loose fill (poured-in). The batts and blankets are cut in sections 15'' or 23'' wide for stud or joist spacing of 16'' o.c. or 24'' o.c. They come with or without a vapor barrier. They are fire and moisture resistant. The loose fill (poured-in) insulation does not have a vapor barrier, and the vapor barrier must be bought and applied separately.

Cellulosic Fiber, Vermiculite and Perlite is loose fill and does not have a vapor barrier. It is best suited for non-standard or irregular joist spacing or when space between joists has many obstructions. For fire resistance check the bags to see if they meet Federal Specifications. If they do they will be clearly labeled.

Rigid Insulation Boards are made from organic fiber, polystyrene foam and urethane. Rigid insulation boards made from organic fiber are used for wall sheathing and serve a dual purpose. Polystyrene and urethane rigid insulation boards have a higher R-value but should be installed strictly according to the manufacturer's recommendation. Some manufacturers use aluminum foil facers for an effective vapor barrier. The standard size is 4' x 8' and their nominal thickness is ⅜'' through 1⅞'' with special order thickness through 2¼''.

Other Factors To Consider:

Vapor barriers prevent water vapors contained in warm air from entering the walls, ceilings and floors. Insulation becomes wet from moisture and will not insulate. Enough moisture will also cause rot in the structure.

Infiltration: In any room containing windows and doors, there is a certain amount of air leakage that seeps in through the cracks, especially

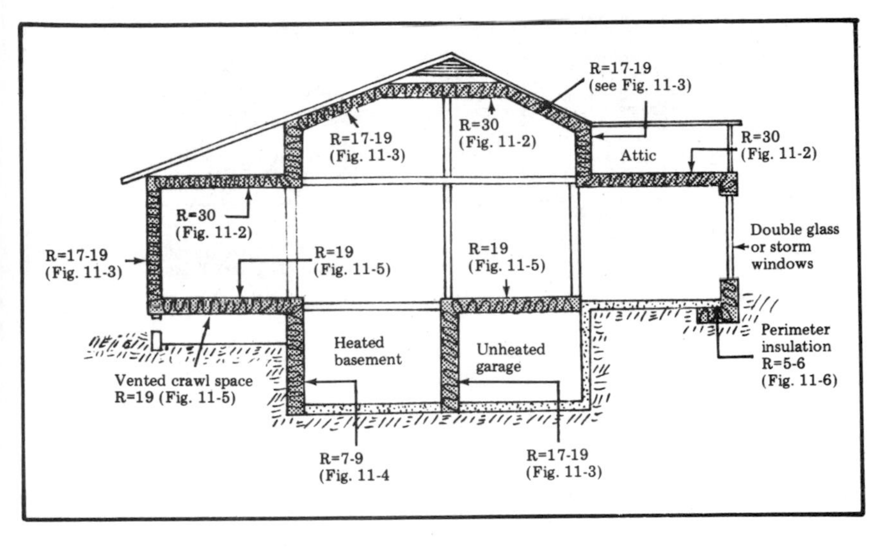

House Insulation
Figure 11-1

when the wind blows. The cold air that leaks into the room imposes an additional load on the heating system. Weatherstripping windows and doors, and sealing around cracks (Figure 11-7) helps prevent infiltration.

Where To Insulate:

Children in an elementary school were asked to share their ideas on how to ease the energy shortage. One child said "when beds aren't in use, take the blankets and put them around the room to help hold the heat." This child had the right idea as to the method of conserving energy. Figure 11-1 shows how a blanket of insulation should be installed around a house for efficient energy conservation. The R-values shown here will vary in different sections of the nation from R-19 for ceilings, R-11 for walls and R-11 for floors in warmer regions to R-38 for ceilings, R-19 for walls and R-22 for floors in the northern or colder regions. Fuel costs in any region are another factor that should not be overlooked in selecting the correct R-value.

Ceilings: Insulation for the ceiling can be either batts, blankets, loose fill (poured-in) or a combination of any two for the desired R-value (Figure 11-2 and 11-10). If the insulation can be installed after the ceiling is finished it will save labor to place the insulation in the attic space before the drywall is installed and install the insulation on the ceiling boards later. If the insulation cannot be installed after the ceiling is finished it will have to be stapled in place between the joists before the drywall is installed. When ceiling insulation is installed *do not block the air flow from the soffit vents to the gable end vents or the roof vents.*

Ceiling Heat Loss Comparison

1. 8" insulation (R-25.3) has 21% less heat loss than 6" insulation (R-19.0).

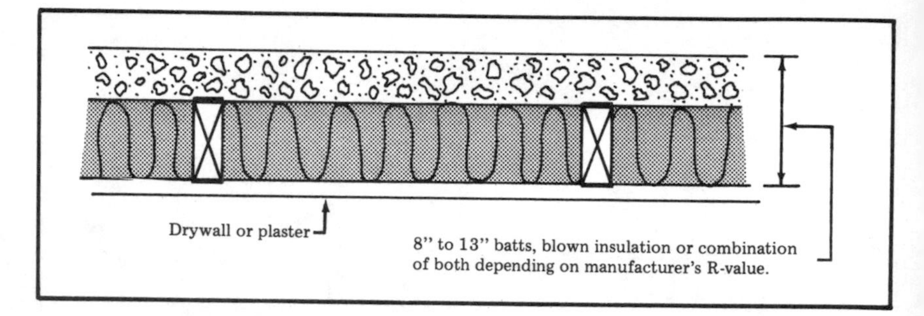

Ceiling Insulation
Figure 11-2

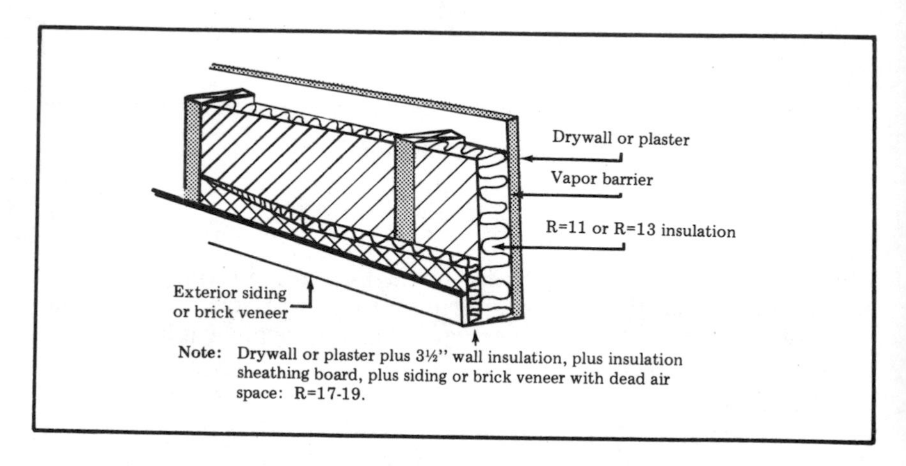

Wall Insulation
Figure 11-3

2. 10" insulation (R-31.7) has 35% less heat loss than 6" insulation (R-19.0).
3. 12" insulation (R-38) has 46% less heat loss than 6" insulation (R-19).

Walls: Exterior walls (excluding windows and doors) account for 13.2% of the total heat loss in a house. All building materials have some insulating value. When R-11 or R-13 batt or blanket insulation (batts or blankets are easier to install than loose fill) are added to the other building material (Figure 11-3) a R = 17-19 value or higher can be attained. Brick veneer with a dead air space can add an extra R-3, or higher. A brick veneer wall with R-11 insulation between the studs and a dead air space has 69% less heat loss than a brick veneer wall with no insulation between the studs.

Masonry walls with no insulation have a very high heat loss, but adding furring strips and R= 7-9 insulation (Figure 11-4) can reduce the heat loss by approximately 83%.

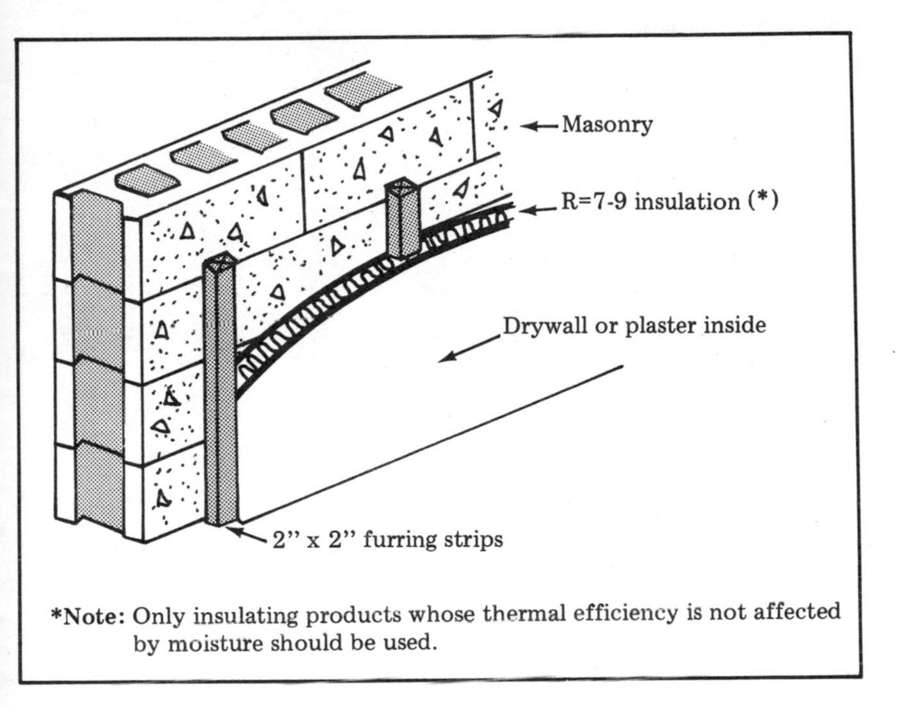

Masonry Wall Insulation
Figure 11-4

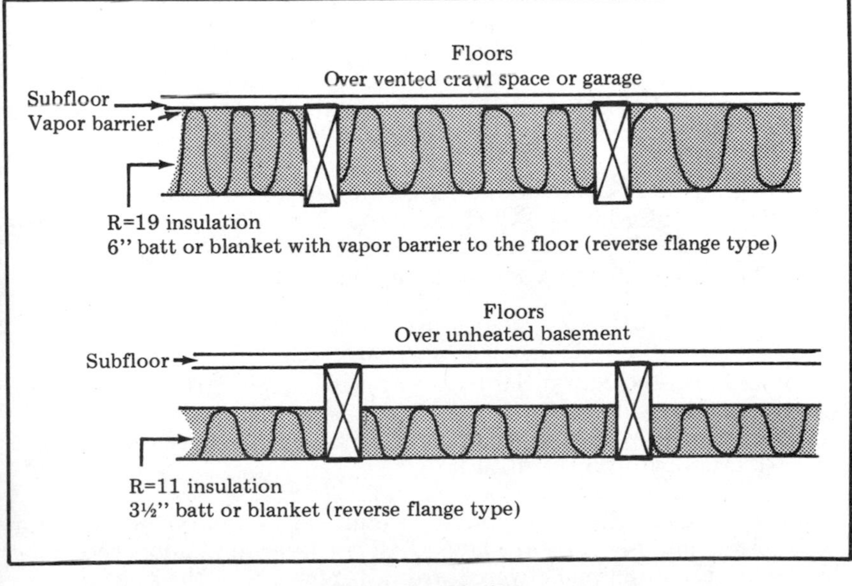

Floor Insulation
Figure 11-5

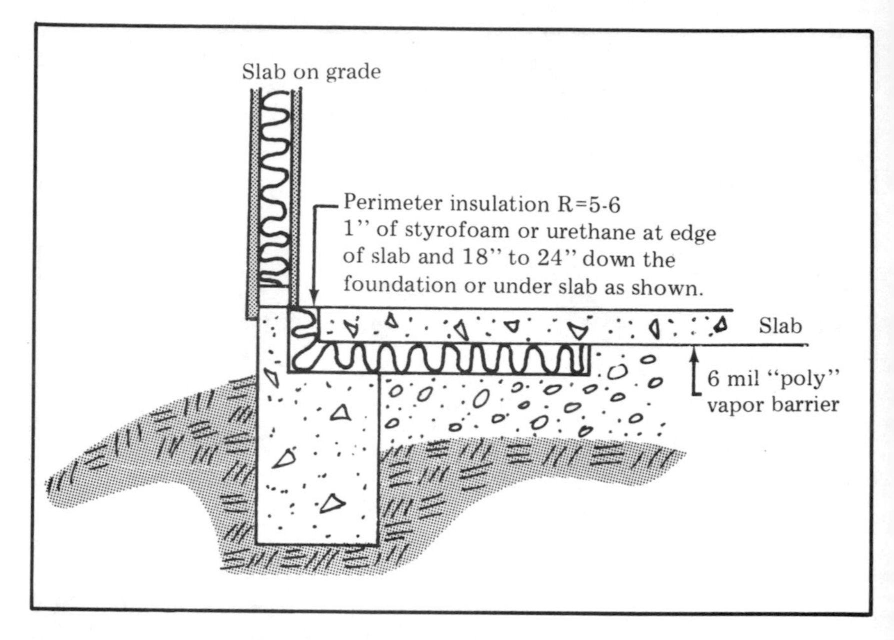

Perimeter Insulation Under Slabs
Figure 11-6

Floors: Floors over heated rooms or basement need no insulation (Figure 11-1). Floors over unheated garages or vented crawl spaces should have insulation with about a R-19 value, depending on the climate (Figures 11-1 and 11-5). Floors over an unheated basement should have R-19 insulation. For slabs on grade as shown in Figures 11-1 and 11-6 perimeter insulation of 1" styrofoam or urethane (R = 5-6) should be installed at edge of slab and 18" to 24" down the foundation or under the slab.

Windows: Windows and sliding glass doors account for 38.1% of the total heat loss in a house. All cracks around windows should be filled with insulation and covered with a vapor barrier as shown in Figure 11-7. Figure 11-11 is a window condensation guide.

Window Heat Loss Comparison
(Windows weatherstripped)

1. Single Pane and Storm Sash have approximately 50% less heat loss than single pane windows with no storm sash.
2. Insulating Glass has approximately 46% less heat loss than single pane windows with no storm sash.

Doors: Outside doors (excluding sliding glass doors) account for 2.4% of the total heat loss in a house. All cracks around doors should be filled with insulation and covered with a vapor barrier the same as windows shown in Figure 11-7.

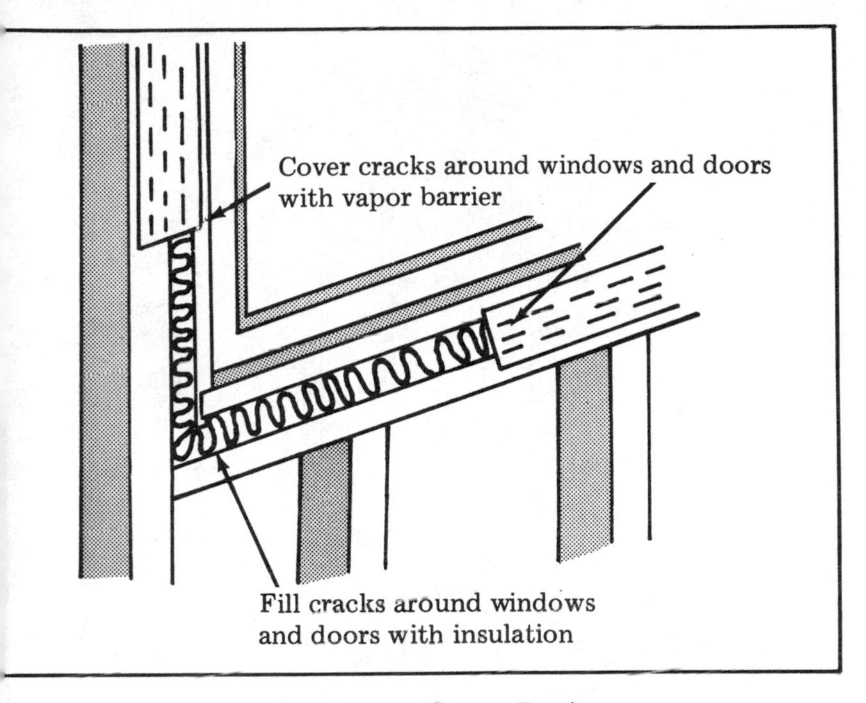

Infiltration and Vapor Barriers
Figure 11-7

Door Heat Loss Comparison
(Doors weatherstripped)

1. Solid core doors (1¾") with storm doors have approximately 35% less heat loss than a solid core door (1¼") with no storm door.
2. An insulated door of urethane/steel with a R-13.8 value (Figure 11-8) with no storm doors has approximately 85% less heat loss than a solid core door (1¾") with no storm door. This same urethane/steel door with no storm door has approximately 76% less heat loss than a solid core door (1¾") with a storm door.

Figure 11-9 shows little spaces that should be insulated to help prevent heat loss and infiltration of air.

The prescribed "R" values can be adversely affected if the intended densities are not maintained. Consequently, the actual effectiveness of insulation is directly affected by the workmanship of the applicator. Compressing batts or blankets changes the density, which in turn changes the "R" value. Blown insulations with too much air in the mixture results in a "fluffy" application with less density than intended. In time settling occurs, and the end result is insufficient thickness to provide the desired "R" value. Some of the more common problems are:

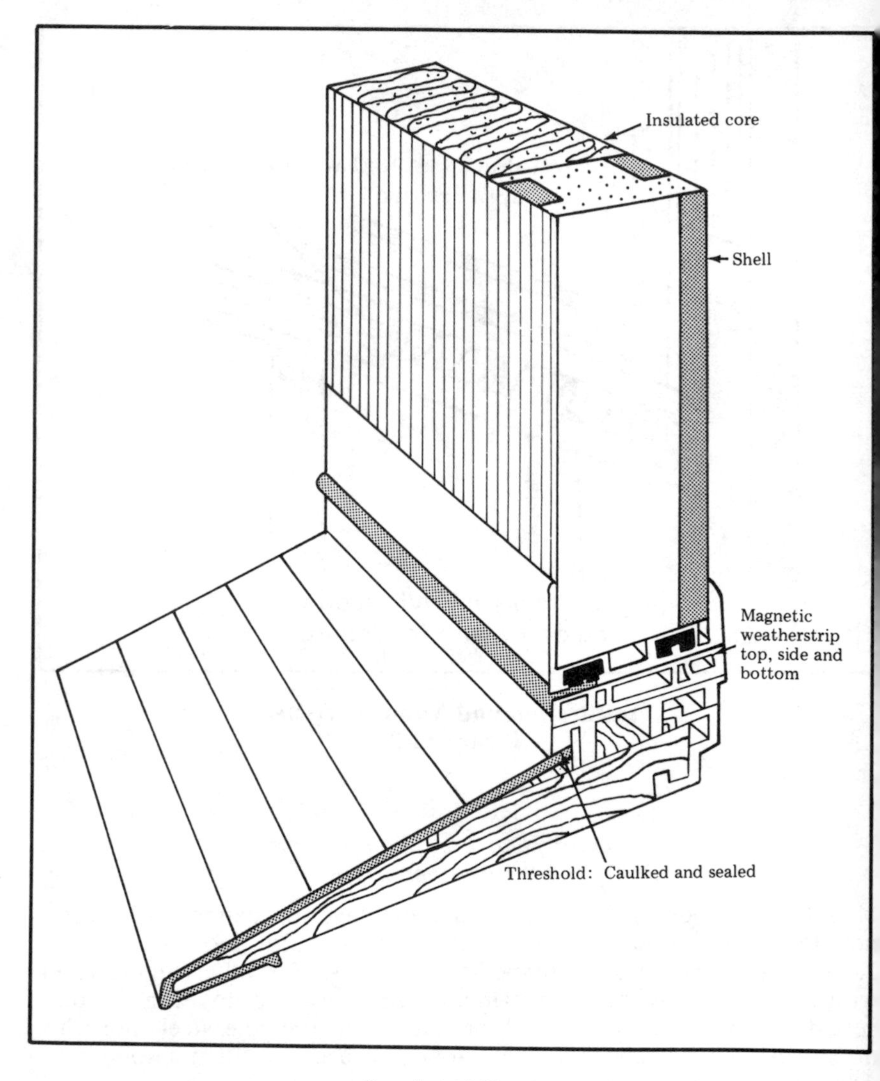

Insulated Door
Figure 11-8

1. Failure to fill small spaces around window and door frames with insulation, and to install a vapor barrier (Figure 11-7).
2. Leaving VOID spaces between framing members as a result of not fitting batts or blankets correctly.
3. Stapling batt or blanket flanges improperly, resulting in a poor vapor barrier.
4. Installing the batt or blanket under floors over crawl spaces with the vapor barrier facing the crawl space instead of the floor.
5. Improper placement and fitting of batts or blankets where cross brac-

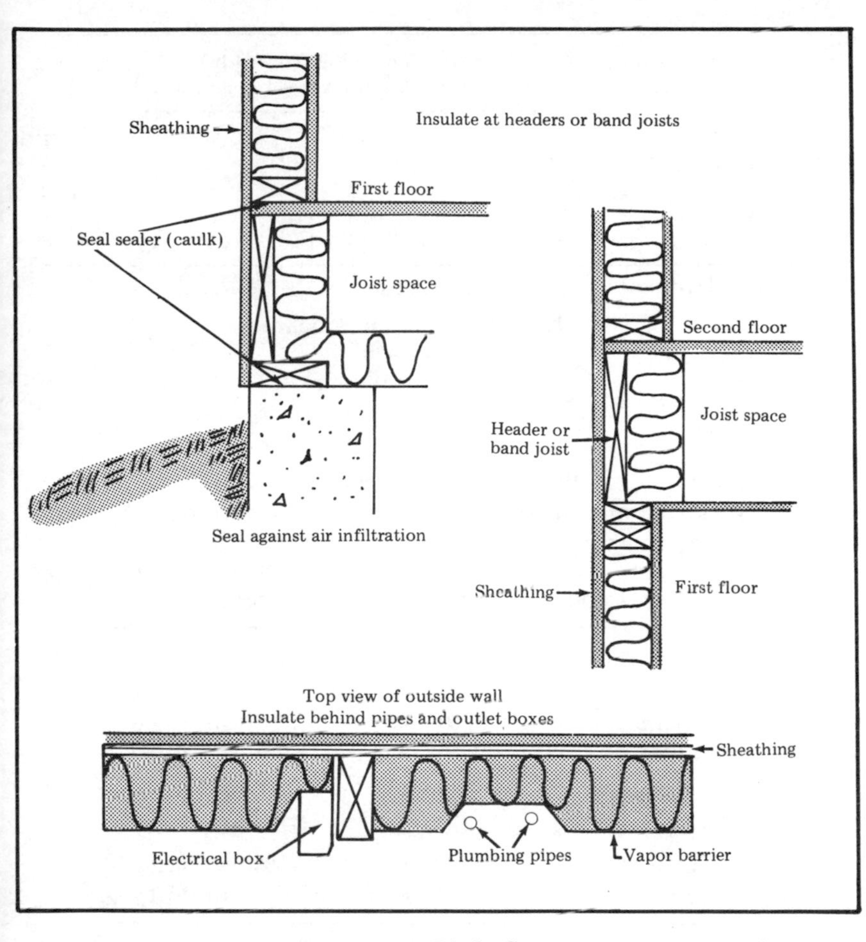

Remember Little Spaces
Figure 11-9

ing (bridging) occurs between joists, and where purlin and/or knee bracing is placed between studs.

Estimating Materials

Take the area of the floor plan for the ceiling and floor insulation; for the walls take the perimeter of the outside walls (plus any interior walls where insulation is required) and multiply by the wall height for the gross area. Deduct any window and door opening 50 square feet or larger for the net area. Divide the area to be insulated by the amount of insulation in each bag or roll and round off to the next higher number for the number of bags or rolls to order. *Example:* 1968 square feet of R-11 insulation (3½" x 15") packed 88 square feet per roll.

$$\frac{1968}{88} = 22.36 \text{ or } 23 \text{ rolls}$$

	Batts or Blankets		Loose Fill (Poured-In)			
	Glass Fiber	Rock Wool	Glass Fiber	Rock Wool	Cellulosic Fiber	
R-11	3½" - 4"	3"	5"	4"	3"	R-11
R-19	6" - 6½"	5¼"	8" - 9"	6" - 7"	5"	R-19
R-22	6½"	6"	10"	7" - 8"	6"	R-22
R-30	9½" - 10½" *	9" *	13" - 14"	10" - 11"	8"	R-30
R-38	12" - 13" *	10½ *	17" - 18"	13" - 14"	10" - 11"	R-38

* Note: Two batts or blankets required

R-Values and Insulation Thickness

Figure 11-10

Window Condensation
(Inside temperature 70° F with low wind velocity)

Window Type	Outside Temperature	Condensation Occurs at:
Single Glass	0° F	12% humidity
	30° F	33% humidity
Insulating Glass	0° F	35% humidity
	30° F	55% humidity
Single Glass with Storm Sash	0° F	40% humidity
	30° F	61% humidity
Insulating Glass with Storm Sash	0° F	53% humidity
	30° F	70% humidity

Figure 11-11

For ceiling insulation where two layers of insulation will be required for the correct R-value, estimate each layer and type separately. *Example:* 1040 square feet for R-30 insulation with trusses spaced 24" o.c. One layer of 6" insulation (R-19) plus one layer of 3½" insulation (R-11) will be combined for the R-30 value.

6" x 23" (R-19): 75 sq. ft. per roll

$$\frac{1040}{75} = 13.87 \text{ or } 14 \text{ rolls}$$

3½" x 23" (R-11): 135.12 sq. ft. per roll

$$\frac{1040}{135.12} = 7.70 \text{ or } 8 \text{ rolls}$$

If the polyethylene film is used as a vapor barrier the width of the film should be the full wall height. Divide the area by the number of square feet per roll and round off to the next highest number for the number of rolls required. *Example:* 1968 square feet of wall area with 8'0" ceilings. Use 8'0" wide polyethylene film (it comes packed 100' in length).

$$\frac{1968}{800} = 2.46 \text{ or } 3 \text{ rolls}$$

The materials to insulate the house in Figures 7-1 and 7-2 are estimated as follows:

1. Ceiling area over Kitchen, Dining Room and Living Room (trusses over are on 24" o.c.) for R-30 insulation..................624 sq. ft.
Ceiling area over second floor (trusses over are on 24" o.c.) for R-30 insulation..1040 sq. ft.
Total area for R-30 insulation..........................1664 sq. ft.

R-19 (6" x 23") insulation: 75 sq. ft. per roll

$$\frac{1664}{75} = 22.19 \text{ or } 23 \text{ rolls}$$

R-11 (3½" x 23") insulation: 135.12 sq. ft. per roll

$$\frac{1664}{135.12} = 12.31 \text{ or } 13 \text{ rolls}$$

For ceilings with 24" o.c. truss spacing for R-30 value.
Order: *23 rolls 6" x 23" (R-19) insulation*
13 rolls 3½" x 23" (R-11) insulation

2. Ceiling over unheated garage and utility area (joists are spaced 16" o.c.) for R-19 insulation...................................624 sq. ft.

R-19 (6" x 15") insulation: 49 sq. ft. per roll

$$\frac{624}{49} = 12.73 \text{ or } 13 \text{ rolls}$$

For ceiling with 16" o.c. joist spacing for R-19 value.
Order: *13 rolls 6" x 15" (R-19) insulation*

3. The total net area of the exterior walls for the first and second floors (studs are spaced 16" o.c.) for R-11 insulation is...........1968 sq. ft.

R-11 (3½" x 15") insulation: 88 sq. ft. per roll

$$\frac{1968}{88} = 22.36 \text{ or } 23 \text{ rolls}$$

For walls with 16" o.c. stud spacing for R-11 value.
Order: *23 rolls 3½" x 15" (R-11) insulation*

4. The floor area over the unheated basement (joists are 16" o.c.) for R-11 insulation (Figure 11-5) is..........................1040 sq. ft.

R-11 3½" x 15") insulation: 88 sq. ft. per roll

$$\frac{1040}{88} = 11.82 \text{ or } \mathit{12\ rolls}$$

For floor with 16" o.c. joist spacing for R-11 value.

Order: *12 rolls 3½" x 15" (R-11) insulation*

Material for stapling insulation:

Order: *2 boxes (5000 per box) ¼" staples*

Polyethylene film:

Order: *None*

(Enter the above material on the Cost Estimate Worksheet For Energy Saving Materials)

Estimating Labor

Past experience on labor costs for installing batt or blanket insulation for ceilings, walls and floors reveals a man-hour factor of .01000 per square foot. Using this factor to estimate the labor costs for installing 7199 square feet of insulation for the house in Figures 7-1 and 7-2 the computation is:

7199 square feet x .01000 factor = 71.99 or *72 man-hours*

Two laborers will be assigned the responsibility of installing this insulation. The pay scale for each workman is $5.00 per hour.

The estimated labor to install 7199 square feet of insulation in the ceilings, walls and floors will be:

72 man-hours @ $5.00 = $360.00

(Enter this labor cost on the Cost Estimate Worksheet For Energy Saving Materials).

Cost Estimate Worksheet For Energy Saving Materials

Insulation:

_________ Rolls/Bags @ _________ = $_________
(Type _________: Size _________)

_________ Rolls/Bags @ _________ = _________
(Type _________: Size _________)

_________ Rolls/Bags @ _________ = _________
(Type _________: Size _________)

_________ Rolls/Bats @ _________ = _________
(Type _________: Size _________)

_________ Rolls/Bags @ _________ = _________
(Type _________: Size _________)

_________ Square feet blown insulation @_________ = _________

_________ Boxes staples (_________M) @_________ = _________
(Size _________)

Storm Windows = _________
(List on separate sheet and enter cost here)

Storm Doors = _________
(List on separate sheet and enter cost here)

Other Material = _________
(List on separate sheet and enter cost here)

$_________ _________

Sales Tax (______%) _________

Cost of material $ _________

Labor $ _________

Total Cost of Labor and Material $ _________

(Enter on Line 11, Form 100)

Chapter 12

Interior Wall and Ceiling Finish

After the rough-in work for the electrical, plumbing, heating and air conditioning has been completed and the wall insulation (and ceiling insulation if necessary at this time) installed, the interior walls and ceilings can be finished. If there are drop ceilings over the kitchen cabinets and bathroom vanities, or other carpentry work to be done before the wall finish is applied, it must be done at this stage of the construction.

Lath and plaster has almost been replaced by thin-coat plaster for plastered walls because of the large amount of water it contains, and the time lost waiting on the plaster to dry. Thin-coat plaster is applied over a plaster base of gypsum core with a special face paper. These plaster base sheets are 4' wide and the normal length used in residential construction is 12'. They are available in ½'' and ⅝'' thickness. The plaster is specially formulated and is applied over the plaster base 1/16'' to 3/32'' thick. The drying time is approximately 24 hours, and it becomes a hard and durable finish. The carpenters can begin the interior trim in the same ''drying'' time as for gypsum drywall construction. The quantity of plaster base is estimated by the square feet the same as for drywall construction (explained later in this chapter). The application of thin-coat plaster is done by plasterers who specialize in this trade. They will contract the job for either the labor and material or the labor only on a per square yard basis. If the builder is to furnish the material, consult the plaster contractor and follow the manufacturer's recommendation as to the type and quantity of material to order.

Drywall, or gypsum wall board, is the most commonly used material for walls and ceilings and it has a high fire rating and comes in standard widths of 4' and lengths from 8' to 16'. The standard thicknesses are ½", ⅝" and ¾", but thicknesses of ¼" and ⅜" are also available for special applications, such as where a radii is required.

To estimate drywall material, calculate the square feet of the walls and ceilings in each room, closet, halls and stairway to be covered. Figure 12-1 is a chart showing the square feet for walls and ceilings for different size rooms with a ceiling height of 8'0". Figure 12-2 is a worksheet for computing the square feet of walls and ceilings for rooms larger, or with a different ceiling height than those shown in Figure 12-1. Sample worksheets of Figure 12-2 are enclosed, including a worksheet for rooms with offsets as shown in Figure 12-3. The approximate weight and bending radii of different size gypsum boards are shown in Figure 12-4. The area is totaled for all areas to be covered. If 4' x 12' boards are to be used (they have less joints to be finished), divide this total area by 48 and round off to the next higher number for the number of pieces to order. If 4' x 8' boards are to be used, divide the total area by 32 and round off to the next higher number for the number of pieces to use. Estimating all walls as solid and not deducting any openings less than 50 square feet will normally take care of the waste factor.

The following allowances can be used as a guideline for the materials to fasten and finish gypsum wallboards.

1. Allow 5 lbs. 1⅜" Annular Ring Nails per 1000 square feet wallboard (used with adhesive).
2. Allow 1 - tube adhesive per 500 square feet wallboard.
3. Allow 1 - roll tape (250') per 600 square feet wallboard.
4. Allow 1 - 5 gal. can joint compound per 1000 square feet wallboard.

NOTE: If ceilings are finished with a texture finish using joint compound, allow 1 - 5 gal. can joint compound per 400 square feet wallboard.

Metal cornerbeads are used on all outside corners. They come in 8'0" lengths. Estimate the number by adding the total lineal feet of the outside corners where they are required, divide by 8 and round off to the next higher number.

The drywall work is normally let on subcontract. The drywall contractor will contract for either all labor and material or for labor only on a per square foot basis.

In the house in Figures 7-1 and 7-2, gypsum drywall boards (½" x 4' x 12') are to be used on all walls and ceilings in the house and garage (except in the family room where prefinished paneling will be used). The material is estimated as follows:

Gypsum boards: The total area of the walls and ceilings where gypsum boards are required is 10,016 square feet.

Length In Feet

Width in Feet

	2'	3'	4'	5'	6'	7'	8'	9'	10'	11'	12'	13'	14'	15'	16'	17'	18'	19'	20'
2'	68	86	104	122	140	158	176	194	212	230	248	266	284	302	320	338	356	374	392
3'	86	105	124	143	162	181	200	219	238	257	276	295	314	333	352	371	390	409	428
4'	104	124	144	164	184	204	224	244	264	284	304	324	344	364	384	404	424	444	464
5'	122	143	164	185	206	227	248	269	290	311	332	353	374	395	416	437	458	479	500
6'	140	162	184	206	228	250	272	294	316	338	360	382	404	426	448	470	492	514	536
7'	158	181	204	227	250	273	296	319	342	365	388	411	434	457	480	503	526	549	572
8'	176	200	224	248	272	296	320	344	368	392	416	440	464	488	512	536	560	584	608
9'	194	219	244	269	294	319	344	369	394	419	444	469	494	519	544	569	594	619	644
10'	212	238	264	290	316	342	368	394	420	446	472	498	524	550	576	602	628	654	680
11'	230	257	284	311	338	365	392	419	446	473	500	527	554	581	608	635	662	689	716
12'	248	276	304	332	360	388	416	444	472	500	528	556	584	612	640	668	696	724	752
13'	266	295	324	353	382	411	440	469	498	527	556	585	614	643	672	701	730	759	788
14'	284	314	344	374	404	434	464	494	524	554	584	614	644	674	704	734	764	794	824
15'	302	333	364	395	426	457	488	519	550	581	612	643	674	705	736	767	798	829	860
16'	320	352	384	416	448	480	512	544	576	608	640	672	704	736	768	800	832	864	896
17'	338	371	404	437	470	503	536	569	602	635	668	701	734	767	800	833	866	899	936
18'	356	390	424	458	492	526	560	594	628	662	696	730	764	798	832	866	900	934	968
19'	374	409	444	479	514	549	584	619	654	689	724	759	794	829	864	899	934	969	1004
20'	392	428	464	500	536	572	608	644	680	716	752	788	824	860	896	932	968	1004	1040

Square Feet of Walls and Ceiling With 8-Foot Ceiling
(No Openings Deducted)
Figure 12-1

Room: ______________				
Size: ________ width x ________ length				
Ceiling Height: ________				
Perimeter of Walls: ____ width + ____ Length x 2 = ____ Lin. ft.				
Area				
Walls: ___ lin. ft. (perimeter) x ___ ceiling height =			________	sq. ft.
Ceiling: ____ width x ____ length =			________	sq. ft.
Total square feet of walls and ceiling			________	sq. ft.*
________ square feet divided by 9 = ________ sq. yds.				
*Note: If openings are to be deducted, deduct them from this total.				

Worksheet for Square Feet of Walls and Ceiling
Figure 12-2

$$\frac{10016 \text{ sq. ft.}}{48} = 208.67 \text{ or } 209 \text{ pieces}$$

Order: *209 pcs. ½" x 4' x 12' gypsum boards*

Nails:

$$\frac{10016 \times 5}{1000} = 50.08 \text{ or } 50 \text{ lbs.}$$

Order: *50 lbs. 1-3/8" annular ring nails*

Adhesive:

$$\frac{10016 \times 1}{500} = 20.03 \text{ or } 20 \text{ tubes}$$

Order: *20 tubes adhesive*

Tape (250' per roll):

$$\frac{10016 \times 1}{600} = 16.69 \text{ or } 17 \text{ rolls}$$

Order: *17 rolls (250') tape*

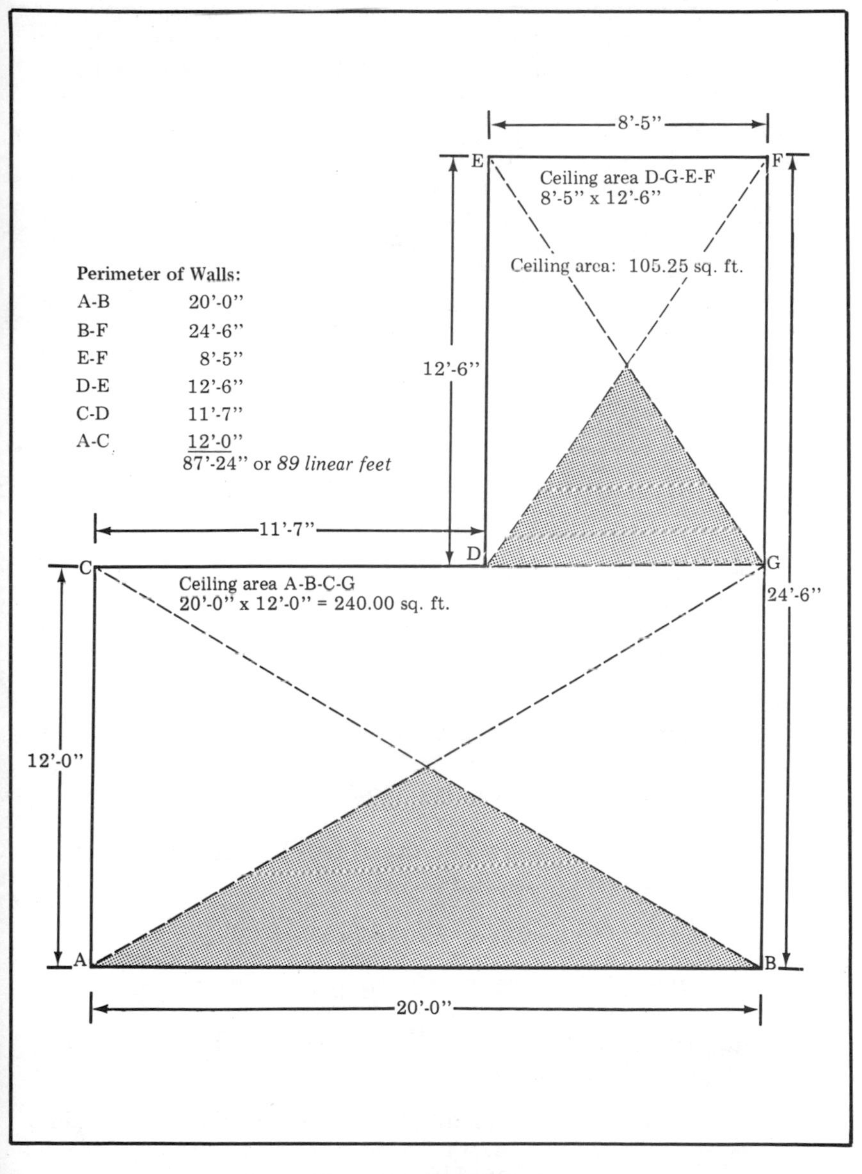

Rooms with Offsets
Figure 12-3

Size	Approximate Weight	Bending Radii Lengthwise	Width
1/4" x 4' x 8'	35 lbs. each	5'	15'
3/8" x 4' x 8'	50 lbs. each	7½'	25'
3/8" x 4' x 12'	75 lbs. each	7½'	25'
1/2" x 4' x 8'	67 lbs. each	10' *	---
1/2" x 4' x 12'	100 lbs. each	10' *	---
5/8" x 4' x 8'	90 lbs. each	---	---
5/8" x 4' x 12'	135 lbs. each	---	---

***Note:** Bending two ¼" pieces succesively permits radii shown for ¼".

Gypsum Board Weight and Bending Radii
Figure 12-4

Joint compound (5-gal. cans):

$$\frac{10016 \times 1}{400\ *} = 25.04 \text{ or } 25\text{-}5 \text{ gal. cans}$$

*Note: *Ceilings are to have a texture finish using joint compound.*

Order: *25-5 gal. cans joint compound*

Metal cornerbeads: There are 196 lineal feet of outside corners where cornerbeads will be required.

$$\frac{196 \text{ lin. ft.}}{8} = 24.50 \text{ or } 25 \text{ pieces}$$

Order: *25 pcs. o/s metal cornerbeads*

(Enter the above material on the Cost Estimate Worksheet for Interior Wall and Ceiling Finish).

Labor: The labor to hang and finish the drywall (including finishing the ceilings with a texture finish) was let on subcontract to a drywall contractor for .15¢ per square foot. This labor cost will be estimated at:

10,016 square feet @ .15¢ = $1502.40

Drop ceilings will be constructed over the kitchen cabinets and bathroom vanities. The carpentry labor to construct these drop ceilings and other coordination work with the drywall contractor is estimated at $175.00.

(Enter these labor costs on the Cost Estimate Worksheet For Interior Wall and Ceiling Finish).

Cost Estimate Worksheet for Interior Wall and Ceiling Finish

Gypsum Wallboard:

_________ pcs. @ _________ = $_________

(Size:_______________)

_________ pcs. @ _________ = _________

(Size: _______________)

_________ pcs. @ _________ = _________

(Size: _______________)

Nails:

_________ lbs. @ _________ = _________

(Size and type: _______________)

Adhesive:

_________ tubes @ ___ = _________

Tape:

_________ rolls @ _________ = _________

(Size of roll: _______________)

Joint Compound:

_________ cans @ _________ = _________

(Size of can: _______________)

Metal Cornerbead:

_________ pcs. @ _________ = _________

Other Material: = _________

(List on separate sheet and enter total cost here)

Sub total $_________ $_________

Sales tax (____%) _________

Cost of material $_________

Labor:

Drywall or plaster labor $_________

Other labor * _________

Cost of labor $_________ $_________

Cost of labor and material $_________

(Enter on line 12: Form 100)

*** Note:** Coordination work with drywall or plaster contractor.

Chapter 13

Exterior Trim

The exterior trim of a house is installed at the same time the electrical, plumbing, heating and air conditioning are roughed-in and the interior walls are being finished. The materials for the exterior trim should be good grade, weather-resistant and able to hold paint to retain their appearance as long as the house stands. The number of joints should be kept to a minimum.

The windows and exterior doors should be in place before the interior walls are finished, but they are estimated with the exterior trim. The materials for the exterior trim listed in this chapter are:

1. Windows, complete with weatherstripping, window locks and window pulls.
2. Exterior doors, complete with door frames, weatherstripping, trim and hardware including door locks.
3. Siding, including accessories and nails.
4. Fascia, frieze and rake boards.
5. Soffit and porch ceilings.
6. Porch column posts.
7. Gable louvers.

8. Garage doors.
9. Shutters.
10. Flashing.

Windows
Refer to the blueprints for the number, type and size of the windows. They may be wood or metal, or a combination of both, double-hung, casement, awning, sliding, fixed or combined. The glass may be single pane with storm windows and screens, or insulated glass with screens that must be bought separately. The plans and specifications will give this information.

Exterior Doors
The blueprints and specifications will specify the number, type and size of the exterior doors. The door frame for the main entrance door may have a special design, and the estimator should not overlook this on the plans and specifications. The doors may be hollow or solid-core or insulated with a metal clad as shown in Figure 11-8. Sliding doors are made of metal or wood and may be single pane or insulated glass. Screens are normally included with sliding doors.

Siding
Siding comes in many styles, shapes, patterns and materials. It may be used as the only cladding on a house, with brick veneer, or a combination of two or more types of siding (see Figure 13-5). Some of the materials used for manufacturing siding are wood, hardboards, aluminum and vinyl. Select a good quality siding that will hold paint. Any money saved by buying cheap siding will be lost when the house's appearance deteriorates soon after. Some of the different types of siding are:

1. Bevel and/or lapped
2. Board and Batten
3. Shiplap
4. Tongue and Groove
5. Panel
6. Wood Shingles

Some siding is designed for horizontal installation, and some for vertical installation.

Hardboard siding is manufactured in siding strips and panels. It is made of wood chips and fiber pressed together to form a hard material. The thickness of lap siding is ⅜'' or 7/16'', and it is available in widths up to 12''; and in lengths from 8' to 16' in multiplies of 2'. The panel siding is ¼'', ⅜'' or 7/16'' thick and the width is 4'. They come in lengths up to 16'.

Aluminum siding is available in vertical V-groove, board and batten and bevel siding. The bevel siding is the most popular. The paint on this siding is baked on and is very durable.

Vinyl siding is a product of modern chemistry. It never needs painting and is easily cleaned with soap and water. It is molded into various textures and shapes, the most popular being the bevel siding.

Wood siding is made from redwood or cedar for their durability, paint holding ability and appearance. Bevel, board and batten, shiplap and tongue and groove sidings are manufactured from these two wood species.

Cedar sidewall shingles are manufactured from select logs with no two shingles being exactly alike. They come in lengths of 16'', 18'' and 24''. One square of shingles will cover 100 square feet with a 5'' exposure for the 16'' length; 100 square feet with a 5½'' exposure for the 18'' length; and 100 square feet with a 7½'' exposure for the 24'' length shingles. Cedar shingles are extremely durable and require no painting or staining.

To estimate siding of any type, first compute the area of the walls, including the gables. Disregard openings less than 50 square feet, thus reducing the waste factor. Refer to the wall sections and floor plans of the blueprints for the dimensions of the walls. After the wall area is computed and the openings deducted, estimate the siding as follows (this allows for lapping and waste):

Wood Bevel Siding (Lapped or Rabbeted)

Nominal Width (Inches)		Multiply Net Wall Area By:
	Wood Bevel Siding (Lapped or Rabbeted)	
8		1.34
10		1.26
12		1.21
	Tongue and Groove	
8		1.37
10		1.33
12		1.31
	Vertical Board Siding	
8		1.10
10		1.08
12		1.07

Aluminum and Vinyl Siding

This siding is sold by the square, which covers 100 square feet on the walls. Divide the net wall area by 100 and add the waste factor recommended by the manufacturer. Round off to the next number for the number of squares required.

Panel Siding
Divide the net wall area by the area covered by one sheet and round off for the number of pieces required. *Example,* 1200 square feet divided by 32 (area of one 4' x 8' sheet) equals 37.50 or 38 pieces.

Cedar Sidewall Shingles
If the recommended exposures, as explained earlier, is used, divide the net wall area by 100 and round off to the next even number of bundles for the number of squares. *Example,* 1265 square feet of wall area is to be covered with 24" cedar shingles with an exposure of 7½". The shingles are packed 4 bundles per square:

$$\frac{1265}{100} = 12.65 \text{ or } 12\tfrac{3}{4} \text{ squares (rounded off to the next whole bundle)}$$

If other than the recommended exposures are to be used, consult the manufacturer's specification sheet for the coverage of one square.

The accessories for the siding such as o/s corners and i/s corners will vary with the type and size of siding used. Consult the manufacturer's specifications for the accessories required.

Fascia, Frieze and Rake Boards
The fascia is a vertical board nailed on the ends of rafters or trusses, and is a part of the cornice (the part of the roof that projects from the wall). The frieze is a vertical board directly under the cornice, or porch ceiling, and adjacent to the wall or beam, see Figures 13-1, 13-2, 13-3 and 13-4. The rake is the sloping edge of a gable roof and the rake board is part of the gable trim, see Figure 13-3.

When estimating fascia, frieze and rake boards determine the size and grade of material from the blueprints and specifications. From the elevations and cornice detail of the blueprints compute the lineal footage (include porches) of each different member listed. Add 5% to the total linear footage of each member for waste and round off to the next multiple of 10. *Example,* the linear footage of 1" x 6" fascia from the prints is 136. Allowing 5% for waste the linear footage is 142.80 (136 + 5%). The material estimated would be 150 linear feet of 1" x 6" fascia boards.

Soffit and Porch Ceilings
The material most commonly used for the soffit and porch ceilings is A-C exterior plywood, because the appearance of only one side is important. The most common thicknesses are ¼", ⅜", ½", ⅝" and ¾". ⅜" x 4' x 8' A-C exterior grade plywood is commonly used for soffits and porch ceilings because it is easy to handle and resists warping. Figures 13-1, 13-2, 13-3 and 13-4 shows soffit and porch ceilings.

When estimating plywood for soffits and porch ceilings, compute the square footage of the area to be covered and divide this total by the number of square feet in each piece of plywood used. Round off to the next number for the number of pieces. *Example,* from the blueprints 296 square feet of area are to be covered with ⅜" x 4' x 8' A-C plywood for

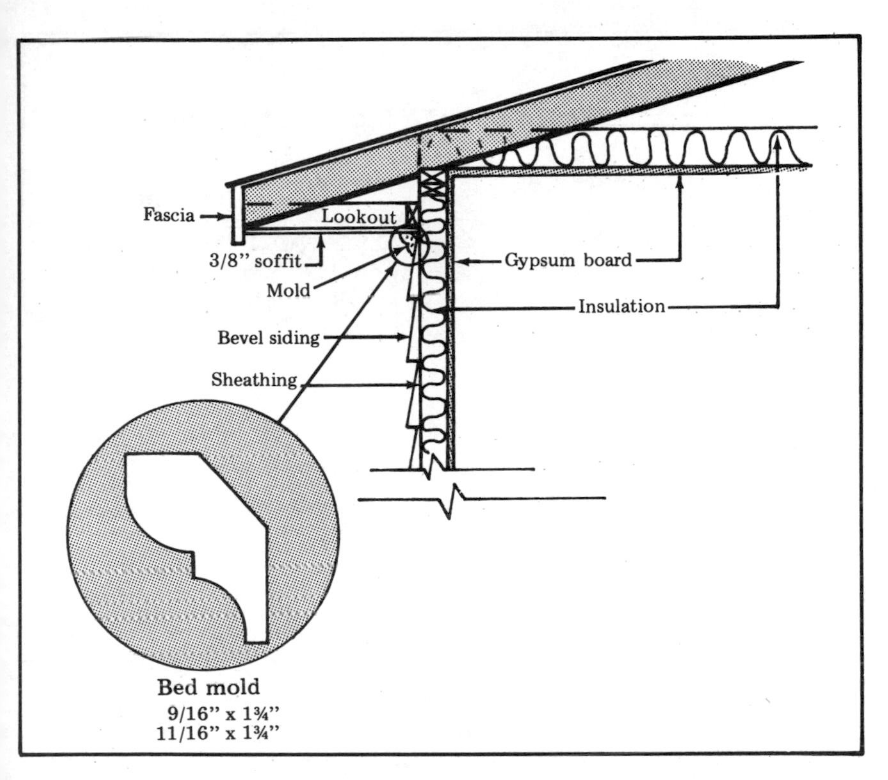

Siding and Trim
Figure 13-1

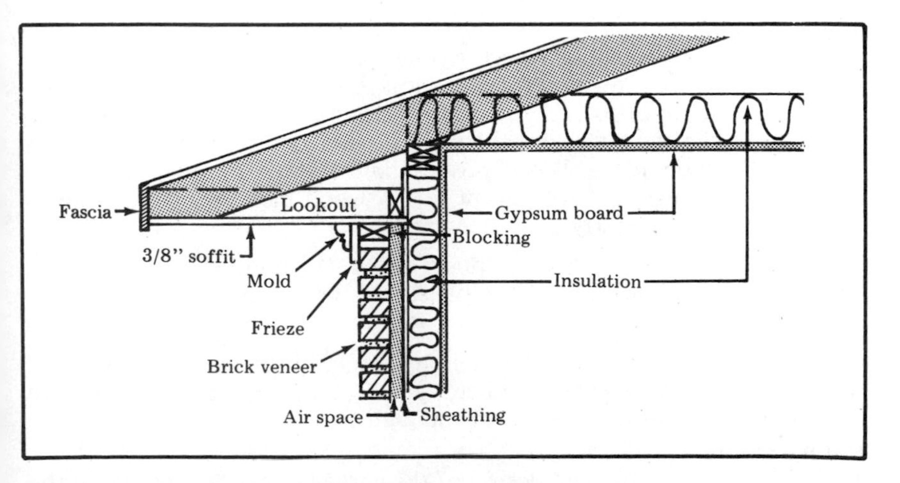

Brick Veneer and Trim
Figure 13-2

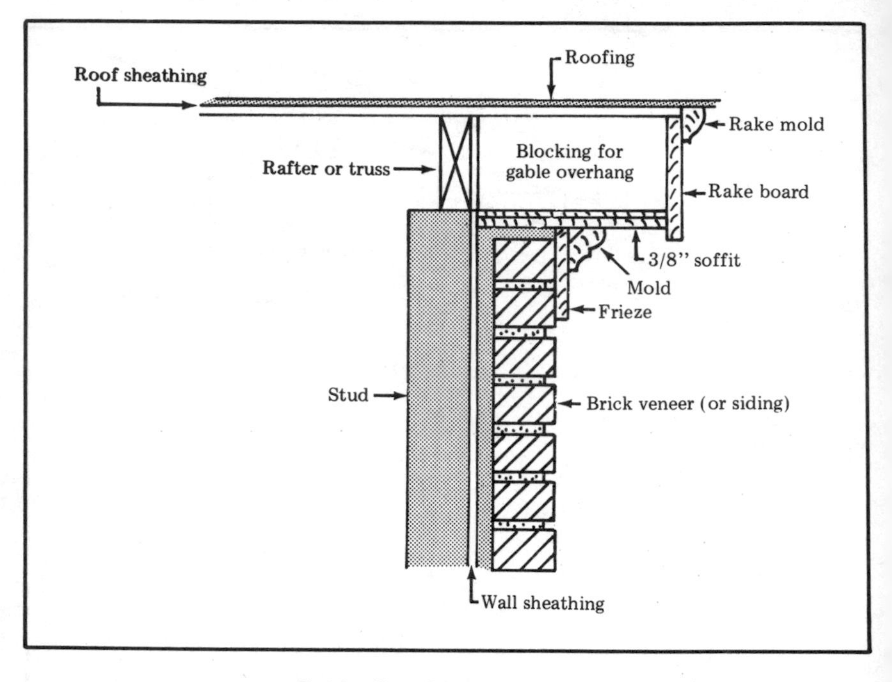

Gable Overhang and Trim
Figure 13-3

the soffits and porch ceilings. 296 square feet divided by 32 (the area of a 4' x 8' sheet) equals 9.25 or 10 pieces of ⅜" x 4' x 8' A-C plywood.

Porch Column Posts
These column posts can be wood, wrought iron or other metal. The type and design will be shown on the elevation section of the plans. The spacing of the column posts is designed by the architect for the load they are to carry, so the span between the posts should never be increased from that shown on the plans. Wood posts should always rest on a metal base as shown in Figure 13-4 to prevent rot. A wood base exposed to dampness will rot very quickly.

When estimating porch column posts obtain the number, type and size from the elevation section of the plans and the specifications.

Gable Louvers
These louvers, or vents, are designed by the architect for proper ventilation and appearance. The number, type and size of gable louvers are shown on the elevation section of the plans.

Garage Doors
The number, size, type and design of the garage doors are shown on the elevation section of the plans. If garage door openers are to be installed they will be listed in the specifications.

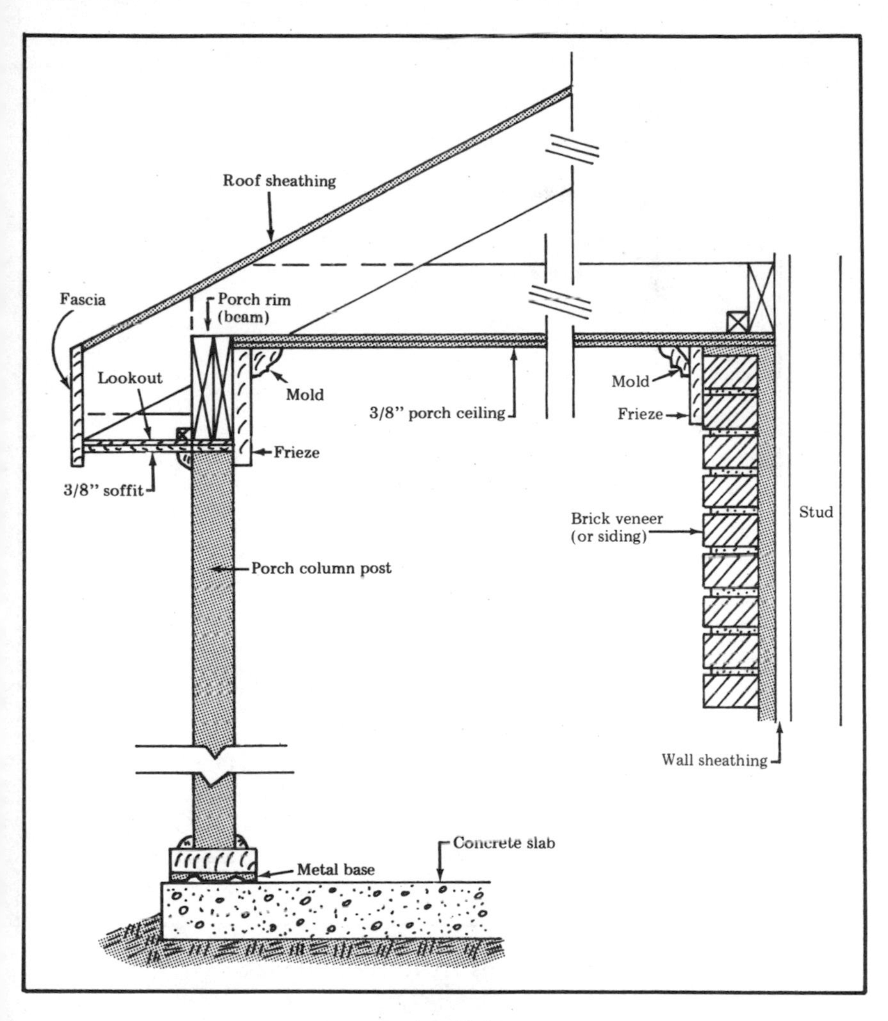

Porch Trim
Figure 13-4

Shutters
The number, size, type and design is obtained from the elevation sections of the plans.

Flashing
Flashing should be provided over all wall openings and intersections as shown in Figure 13-5, and it should be corrosion-resistant sheet metal.

When estimating metal flashing in rolls allow 7" for the width of all wall openings and intersections. From the elevation sections of the plans take the lineal feet of all areas where flashing is required, and if 14"

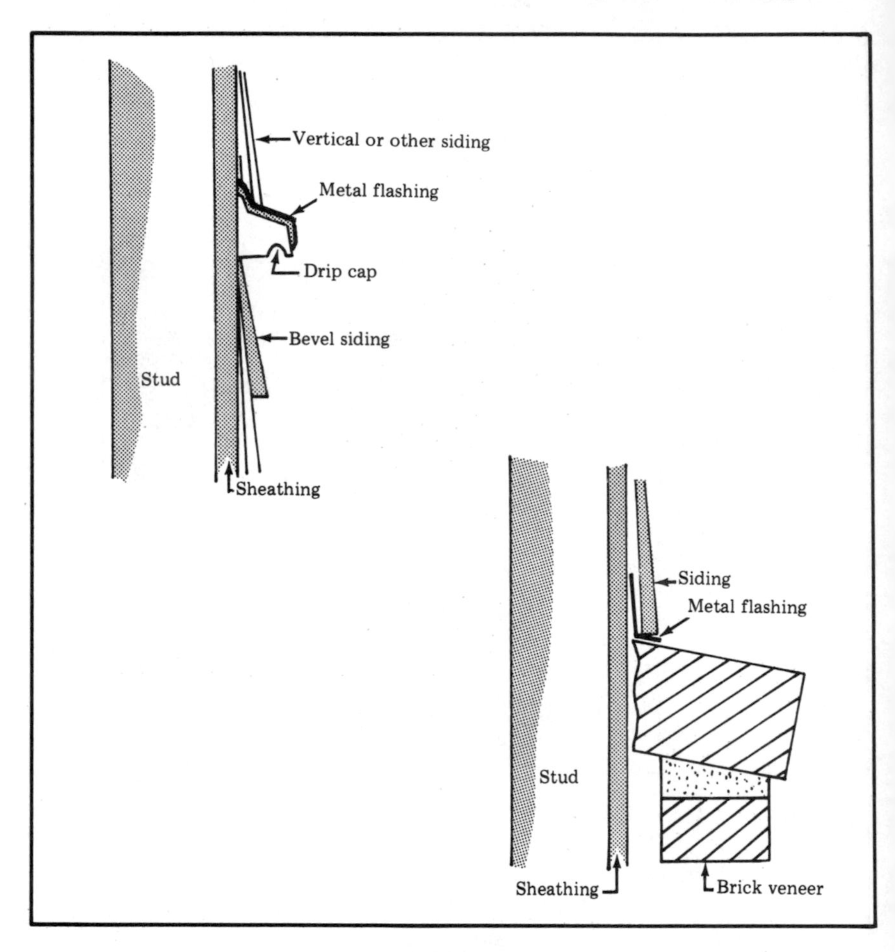

Intersection of Different Materials
Figure 13-5

aluminum flashing is used divide the total lineal feet by 2. *Example,* from the plans 100 lineal feet of flashing will be required for the wall openings and intersections. 14" aluminum flashing will be used, so the amount needed is 100 lineal feet divided by 2, equals 50 lineal feet of 14" aluminum flashing.

Figures 13-6 and 13-7 show different types of molds and their sizes that are commonly used for exterior trim. They are estimated by the lineal feet. Allow 5% for waste.

The nails for exterior trim should be rust resistant and may be estimated as follows:

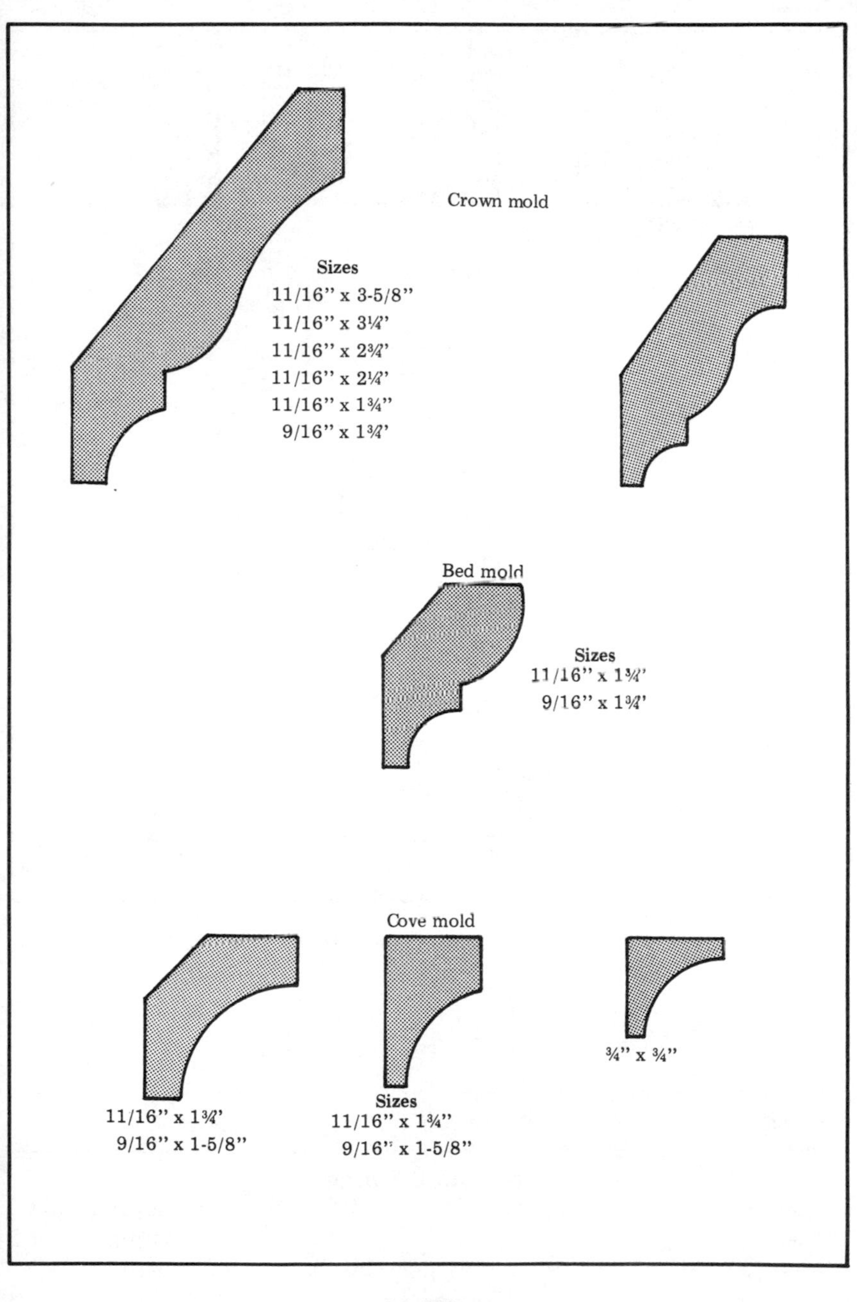

Molding
Figure 13-6

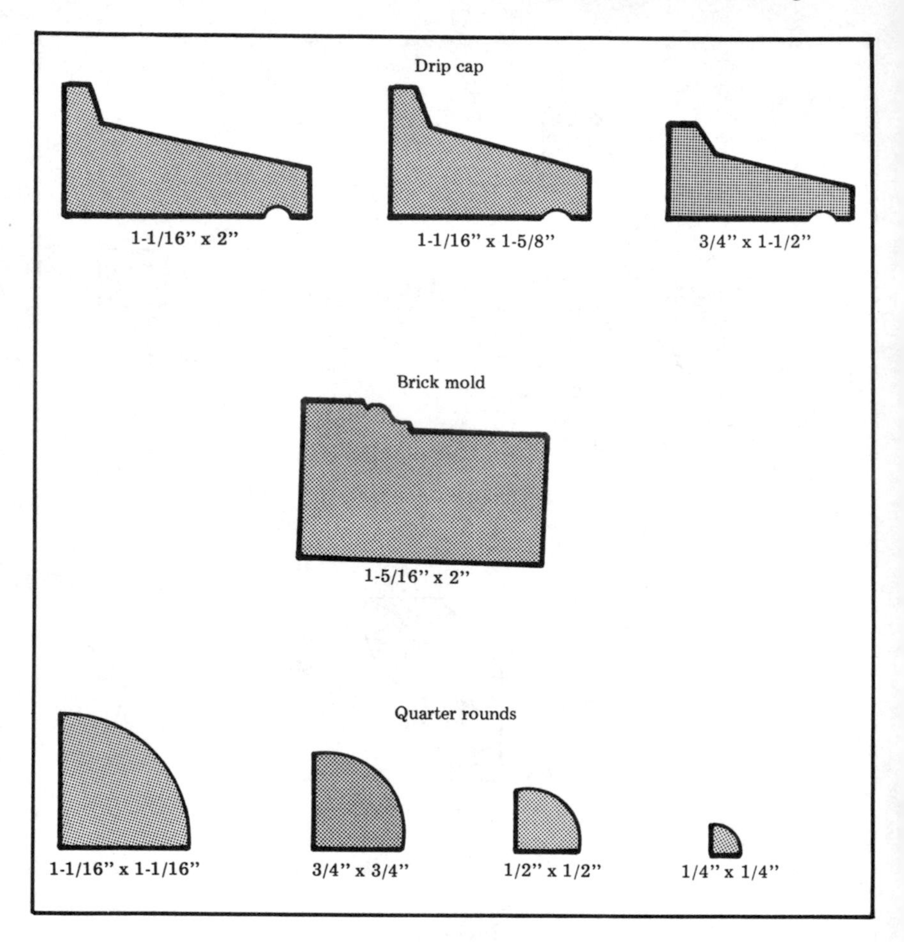

Molding
Figure 13-7

Nail Size	Allow
Horizontal Siding	
6d	6 lbs. per 1000 square feet
8d	7 lbs. per 1000 square feet
10d	10 lbs. per 1000 square feet
Panel Siding	
6d	17 lbs. per 1000 square feet
8d	21 lbs. per 1000 square feet
10d	30 lbs. per 1000 square feet

Nail Sizes	Allow
Cedar Shake Siding	
6d	1 lb. per square
8d	1½ lbs. per square
Soffit and Porch Ceiling	
4d	1 lb. per 100 square feet
6d	1¼ lbs. per 100 square feet
Cornice	
6d or 8d	1 lb. per 100 linear feet

The labor for siding can vary from 2 man-hours per 100 square feet for panel siding to 6 man-hours for cedar shake siding. Another factor that should not be overlooked in the labor output is the height of the walls. Erecting and dismantling scaffolding is dead time, but it must be included in the labor costs. Here again the most accurate labor estimates are always from previous job records. Allow 2 man-hours for setting each window and door frame. For the fascia allow 14 man-hours per 100 lineal feet; for the rake allow 16 man-hours per 100 lineal feet. For porch ceilings including the frieze boards and moldings allow 10 man-hours per 100 square feet.

The material for the exterior trim for the house in Figures 7-1, 7-2 and 7-10 is estimated as follows:

Windows

All windows from the first floor, including weatherstripping, insulated glass, window screens, and all window trim is included in the package price of the house. The windows (except the picture windows) are installed at the factory. No storm windows are needed.

Exterior Doors

All exterior doors except the basement door, are included in the house package. They are pre-hung and include weatherstripping, hardware, locks and trim. The double main entrance doors and the service door leading from the family room to the terrace are insulated doors with metal cladding as shown in Figure 11-8. No storm doors are used. The price of the pre-hung basement door including weatherstripping, hardware, lock and trim is *$117.54.*

Siding

All walls and gables on the second floor, plus the front and back porch gables, will have ⅝'' x 4' x 8' panel siding. The total area of walls to be covered is 1176 square feet. The number of panels that will be required is estimated as follows:

$$\frac{1176 \text{ sq. ft.}}{32} = 36.75 \text{ or } 37 \text{ pieces}$$

37 pcs. ⅝'' x 4' x 8' panel siding is included in the house package.

Fascia, Frieze and Rake Boards
There are 193 lineal feet for the fascia boards. Allowing 5% for waste, the total is 202.65 lineal feet (193 + 9.65 = 202.65). Rounding off to the next multiple of 10 gives 210 lineal feet. 210 lineal feet of 1'' x 6'' fascia boards will be required.

There are 176 lineal feet for the frieze boards. Allowing 5% for waste the total is 184.80 lineal feet (176 + 8.80 = 184.80). Rounding off to the next multiple of 10 gives 190 lineal feet. 190 lineal feet of 1'' x 6'' frieze boards will be needed.

There are 137 lineal feet for the rake boards. Allowing 5% for waste the total is 143.85 lineal feet (137 + 6.85 = 143.85). Rounding off to the next multiple of 10 gives 150 lineal feet. 150 lineal feet of 1'' x 6'' rake boards will be needed.

The fascia, frieze and rake boards are all 1'' x 6'', therefore the combined order will be:

Fascia boards	210 linear feet
Frieze boards	190 linear feet
Rake boards	150 linear feet
Total -	550 linear feet

All of this material is included in the house package.

Soffit and Porch Ceilings
⅜'' x 4' x 8' A-C exterior plywood will be used for the soffit and proch ceilings. There are 804 square feet to be covered. The number of pieces required will be:

$$\frac{804 \text{ sq. ft.}}{32} = 25.13 \text{ or } 26 \text{ pieces}$$

Quantity needed: 26 pcs. ⅜'' x 4' x 8' A-C exterior plywood
This material is included in the package price of the house.

Molding
The mold shown between the soffit and siding in Figure 13-1; between the soffit and frieze board in Figures 13-2 and 13-3; and porch ceiling and frieze boards in Figure 13-4 is 11/16'' x 1¾'' bed mold, as shown in Figure 13-6. The lineal feet is 360 and adding 5% for waste makes the total 378 (360 + 18 = 378). Rounding off is 380 lineal feet.

Quantity needed: 380 lineal feet 11/16'' x 1¾'' bed mold.

A drip cap will be used for the horizontal joint at the gables where the siding is used, see Figure 13-5. The lineal feet is 96. No allowance for waste is necessary. Rounding off, the total is 100 lineal feet.

Quantity needed: 100 lineal feet ¾'' x 1½'' drip cap (see Figure 13-7)

There are 136 lineal feet for the rake mold as shown in Figure 13-3. Rounding off, the total is 140 lineal feet (no allowance is made for waste).

Quantity needed: 140 lineal feet rake mold.

All of the molding is included in the house package.

Porch Column Posts
These posts, trim and metal bases (Figure 13-4) are included in the house package as shown on the blueprints and specifications.

Gable Louvers
These louvers are included in the house package as shown on the plans.

Garage Doors
There are two 9/0 x 7/0 garage doors. These doors are supplied by a local overhead garage door dealer for $460.00 installed. No garage door openers are specified.

Flashing
The flashing for the openings and intersections as shown in Figure 13-5 is included in the house package. It is in pre-formed strips 10'0'' in length. The lineal feet for the openings and intersections where the siding is used is 91.

Quantity needed: 100 lineal feet flashing.

Shutters
The shutters and screws to install them are included in the house package as shown on the plans and specifications.

Nails
The rust-resistant nails for all the exterior trim is included in the house package and the quantity is estimated as follows:

Panel Siding (8d nails)

Note: There are 1184 square feet of panel siding and 21 lbs. nails per 1000 square feet will be the estimating factor.

$$\frac{1184 \text{ sq. ft.}}{1000} \times 21 = 24.86 \text{ or } \mathbf{25\ lbs.}$$

Cornice (8d nails)

Note: There are a total of 550 lineal feet of fascia, frieze and rake boards combined. Allow 1 lb. nails per 100 lineal feet for the factor.

$$\frac{550}{100} \times 1 = 5.50 \text{ or } \mathbf{6\ lbs.}$$

Soffit and Porch Ceiling (6d nails)

Note: There are 832 square feet of soffit and porch ceiling material. The nail factor is 1¼ lbs. per 100 square feet.

$$\frac{832}{100} \times 1\tfrac{1}{4} = 10.40 \text{ or } \mathbf{11\ lbs.}$$

An additional allowance of 8d and 10d nails is included for installing the picture window and door frames, porch column posts and gable louvers.

(Enter all of the above material on the Cost Estimate Worksheet)

Labor
Past job records show that a man-hour factor of 0.14025 for the first floor with very little scaffolding, and a factor of 0.18700 for the second floor can be multiplied by the total square feet of the siding plus the total square feet of the soffit and porch ceilings for a reasonably accurate man-hour estimate for the exterior trim. *Example,* the total square feet of the siding in this estimate is 1184 and the total square feet of the soffit and porch ceiling is 832.

Siding	1184 square feet
Soffit and porch ceiling	832 square feet
Total -	2016 square feet

Most of the work is on the second floor which requires more scaffolding, therefore the man-hour factor of 0.18700 will be used. The man-hours estimated for the exterior trim including the erection and dismantling of the scaffolding is:

0.18700 (factor) x 2016 (sq. ft.) = 377 man-hours*
*Note: Cold weather can increase this time from 10% to 14%.

One carpenter at $8.50 per hour, one carpenter helper at $6.25 per hour, and one laborer at $5.00 per hour will be assigned to the exterior trim. The man-hours for each workman will be estimated as follows:

$$\frac{\text{Total man-hours}}{\text{Number of workmen}} \quad \frac{377}{3} = 125.67 \text{ or } \mathbf{126}\ \textit{man-hours each}$$

The cost of the labor for the exterior trim will be estimated as follows:

126 hours @ $8.50 =	$1071.00
126 hours @ $6.25 =	787.50
126 hours @ $5.00 =	630.00
Total	$2488.50

(Enter this estimated labor cost on the Cost Estimate Worksheet)

Cost Estimate Worksheet For Exterior Trim

Windows: $______
(List on separate sheet and enter total cost here)

Exterior Doors: ______
(List on separate sheet and enter total cost here)

Siding:
_____ square feet/pieces @ ______ = ______
(Type ______: Size ______)

Accessories (o/s and i/s corners, etc.) ______

Fascia, Frieze and Rake Boards:
_____ linear feet @ ______ = ______
(Type ______: Size ______)
_____ linear feet @ ______ = ______
(Type ______: Size ______)
_____ linear feet @ ______ = ______
(Type ______: Size ______)

Soffit and Porch Ceiling:
_____ square feet/pieces @ ______ = ______
(Type ______: Size ______)
_____ square feet/pieces @ ______ = ______
(Type ______: Size ______)
_____ square feet/pieces @ ______ = ______

Molding:
_____ linear feet @ ______ = ______
(Type ______: Size ______)
_____ linear feet @ ______ = ______
(Type ______: Size ______)
_____ linear feet @ ______ = ______
(Type ______: Size ______)
_____ linear feet @ ______ = ______
(Type ______: Size ______)

Porch Column Posts:
_____ number @ _____ = ______
(Type ______: Size ______)
_____ number @ _____ = ______
(Type ______: Size ______)

Gable Louvers:
_____ number @ _____ = ______
_____ number @ _____ = ______

Garage Doors: (with/without garage door openers)
_____ number @ _____ = ______
(Type ______: Size ______)

Flashing:
_____ linear feet @ _____ = ______
(Type ______: Size ______)

Shutters:
_____ number @ _____ = ______
(Type ______: Size ______)

Nails:
_____ lbs. @ _____ = ______
(Size ______)
_____ lbs. @ _____ = ______
(Size ______)
_____ lbs. @ _____ = ______
(Size ______)

Other Material: ______
(List on separate sheet and enter total cost here)

Sub total ______ $______
Sales tax (_____%) ______ $______

Cost of Material $______

Labor: $______

Cost of Labor and Material $______

(Enter on Line 13, Form 100)

Chapter 14

Concrete Floors, Walks and Terrace

The versatility and durability of cement are phenomenal. Cementing materials have been used since the dawn of civilization. Many aqueducts and other cement-bonded structures built by the Romans are still in good condition. However, little was known about the chemistry of these materials until about the middle of the eighteenth century. Portland cement derives its name from a patent that was taken out in 1824 in England for the manufacture of an improved cement, which hardened into a yellowish-gray mass resembling stone found in various quarries on the Isle of Portland, England.

Concrete is a mixture of fine and coarse aggregates surrounded and held together by hardened Portland cement paste. Since these materials are inorganic, concrete is impervious to decay, fire, termites, and rodents.

There is a direct link between the strength of concrete and the proportion of water. If too much water is added, the paste becomes thin and will be weak and porous when it hardens. Cement paste made with the correct amount of water has strong binding qualities to hold the particles of aggregate firmly together to make a strong, dense, watertight concrete. Six gallons of water per sack of cement is the recommended proportion.

Pouring concrete in residential construction has to be coordinated with the carpentry work. For example, the garage floor should be poured before the garage doors are installed; the porch floor must be poured before the porch column posts and railing can be installed; the basement floor should be poured before the basement stairs are constructed and the outside basement door and trim installed.

Concrete weighs approximately 4000 lbs. per cubic yard. Concrete for residential construction is more likely to be delivered to the job site by truck. Estimating concrete is fairly simple, and is done by the cubic yard. Cubic yards are computed by multiplying the length times the width times the depth (in decimal equivalents of a foot) and dividing by 27. *Example:* a garage is 24' x 26' (inside measurement) and is to have an average of 4½'' concrete poured for the floor. The cubic yards required are computed as follows:

$$\frac{24' \times 26' \times .375'\ (4\tfrac{1}{2}'')}{27} = 8.67 \text{ cubic yards}$$

Use Figures 14-1 and 14-2 to speed these computations. *Example:* using the same garage dimensions and concrete depth above, the solution using the factors in Figure 14-2 is:

Area 624 sq. ft. (24' x 26') x factor .01389 (4½'') = 8.67 cubic yards.

The compression strength of concrete is measured in pounds per square inch (psi). A 6 bag mix (6 sacks of cement per one cubic yard) is rated at 3000 psi. A 5 bag mix would be approximately 2500 psi. All concrete exposed to traffic and weather should have a compression strength of at least 3000 psi.

Crushed stone

When concrete is poured in a closed area such as a house floor, basement or garage floor that is subject to dampness, 4'' of stone should be placed under the concrete slab to help prevent hydrostatic pressure (Figure 14-8). Crushed stone is sold by the cubic yard or by the ton. To estimate the tons of crushed stone required, first compute the cubic yards. To convert cubic yards to tons (crushed stone weighs approximately 2700 lbs. per cubic yard), either multiply the cubic yards by 2700 lbs. and divide by 2000 (2000 lbs. per ton), or multiply the cubic yards by the factor of 1.35 (Figure 14.2). *Example:* a basement floor 42' x 28' is to have 4'' of crushed stone under the concrete slab. The cubic yards will be:

Area in square feet times factor for 4'' from Figure 14-2

1176 sq. ft. (42' x 28') x .01235 = 14.52

To convert cubic yards to tons:

14.52 cu. yds. x 1.35 = 19.60 tons

Proof: $$\frac{14.52 \text{ cu. yds.} \times 2700 \text{ lbs.}}{2000} = 19.60 \text{ tons}$$

Cracking in concrete is caused by stresses that exceed the tensile strength of the slab. Highest on the list of these forces is drying shrinkage and drastic temperature changes early in the life of the concrete. All concrete shrinks on drying and all the materials used in concrete affect its shrinkage in some way. But the water content is by far the most impor-

Area In Square Feet	Thickness in inches and decimal equivalents of a foot 3" .250'	3½" .292'	4" .333'	4½" .375'	5" .417'
05	.046	.054	.062	.069	.077
10	.093	.108	.123	.139	.154
20	.185	.216	.247	.278	.309
30	.278	.324	.370	.417	.463
40	.370	.433	.493	.556	.618
50	.463	.541	.617	.694	.772
60	.556	.649	.740	.833	.927
70	.648	.757	.863	.972	1.08
80	.741	.865	.987	1.11	1.24
90	.833	.973	1.11	1.25	1.39
100	.926	1.08	1.24	1.39	1.54
200	1.85	2.16	2.47	2.78	3.09
300	2.78	3.24	3.70	4.17	4.63
400	3.70	4.33	4.93	5.56	6.18
500	4.63	5.41	6.17	6.94	7.72
600	5.56	6.49	7.40	8.33	9.27
700	6.48	7.57	8.63	9.72	10.81
800	7.41	8.65	9.87	11.11	12.36
900	8.33	9.73	11.10	12.50	13.90
1000	9.26	10.81	12.33	13.89	15.44

Cubic yards = sq. ft. x thickness (in decimal equivalents of a foot) divided by 27. (Example, 285 sq. ft. x .375' (4½") divided by 27 = 3.96 cu. yds.)

Check: From above tables	200 sq. ft. (4½" thick)	= 2.78 cu. yds.
	80 sq. ft.	= 1.11 cu. yds.
	05 sq. ft.	= 0.069 cu.yds.
Total	285 sq. ft.	3.959 cu.yds.

Cubic Yard Contents
Figure 14-1

tant factor. If the concrete is free to move as it shrinks, no cracking will occur. But if there is any restraint at the edges of a slab, the concrete may crack.

Cracking is controlled in slabs by providing carefully placed expansion joints that are free to open and close. The most common joint materials are made from asphalt, fiber, and asphalt impregnated fiber. They are available in various thicknesses of ¼", ⅜", ½", ¾" and 1", and in widths from 2" to 8". Sheets are available to be cut as needed. The quan-

Thickness in Inches						
3"	3½"	4"	4½"	5"	5½"	6"
.00926	.01080	.01235	.01389	.01543	.01698	.01852

Area In Square Feet x Factor For Thickness = Cubic Yards

Example: 695 square feet @ 4" thickness
695 sq. ft. x .01235 factor = 8.58 cubic yards

Example: 1065 square feet @ 4½" thickness
1065 sq. ft. x .01389 factor = 14.79 cubic yards

Crushed Stone

Cubic yards multiplied by 1.35 equals tons of crushed stone.*
*Based on one cubic yard of crushed stone weighing 2700 lbs.

Example: 8.58 cubic yards x 1.35 = 11.58 tons crushed stone
14.79 cubic yards x 1.35 = 19.97 tons crushed stone

Multiplying Factors for Cubic Yards
Figure 14-2

tity is estimated in linear feet. Allow 1 man-hour per 100 linear feet to install the expansion joints.

Welded wire fabric or steel bars are used in slabs on grade to control cracking, but not to prevent it. If a reinforced slab cracks, the steel will hold the cracks together so that loads can be transferred across the face of a crack by the interlocked aggregate.

Welded wire fabric
Welded wire fabric or wire mesh is sold by the roll, a roll being 5' x 150' (750 square feet). It is designated by the spacing and gauge of the wire. *Example:* wire mesh designated as 6 x 6: #10, or 6 x 6: 10/10 means the longitudinal wires are spaced 6" o.c. and the transverse wires are spaced 6" o.c. Both the longitudinal and transverse wires are 10 gauge. The mesh may be square or rectangular, such as 4 x 8: 8/12, and so on. In this example, 4" is the longitudinal spacing, 8" the transverse spacing. 8/12 designates the gauge of the wire as #8 gauge for the longitudinal and #12 gauge for the transverse wire.

Reinforcing rods
Reinforcing rods are designated by numbers (Figure 14-5). When a slab is poured over a fill, reinforcing rods are recommended because they are stronger than wire mesh, and can be supported by a masonry wall. When a walk adjacent to a house or a porch is poured over a fill, a grade beam supported by piers that extend down to solid earth may be required for the supporting base for the reinforcing rods (Figure 14-6). The blueprints may not show this grade beam, but the estimator should not overlook it in his estimate. Figures 14-3 and 14-4 are forms and a diagram showing how to estimate reinforcing rods.

Horizontal Rods and Transverse Rods @ 16" x 16"

Width ____ x .75 less 1 = ____ x length ____ = ____ lineal feet

Length ____ x .75 less 1 = ____ x width ____ = ____ lineal feet

____ lineal feet

Allow (____%) for overlapping ____

Total ____ lineal feet

Horizontal Rods and Transverse Rods @ 12" x 12"

Width ______ less 1 = ______ x length ______ = ______ lineal feet

Length ______ less 1 = ______ x width ______ = ______ lineal feet

______ lineal feet

Allow (____ %) for overlapping ______

Total ______ lineal feet

Estimating Reinforcing Rods
Figure 14-3

Tie wire
Tie wire for reinforcing rods is sold by the pound. Allow 1 lb. per 400 lineal feet of rods.

Vapor barrier
Polyethylene film (4 or 6 mil) is used as a vapor barrier under concrete and it is available in various widths up to 20' in rolls 100' in length. It is placed over the ground or stone and immediately under the concrete slab. Allow 5% for overlapping.

Walks that exceed a grade of 5% (5/8'' per foot) in areas subject to freezing should be constructed with ramp and risers (Figure 14-7).

Forms and Screeds
Forms must support the concrete against hydrostatic pressure until it has hardened. They are used for slabs on porches, garage door openings, walks, terraces and patios. They should be rigid enough to support the concrete they are to hold, well braced with stakes (2'' x 2'' x 18'') to prevent displacement,and tight enough to prevent concrete leakage. When curves are formed (such as round patios and curved walks), use plywood or hardboard that can be bent to the proper radii, and brace them well.

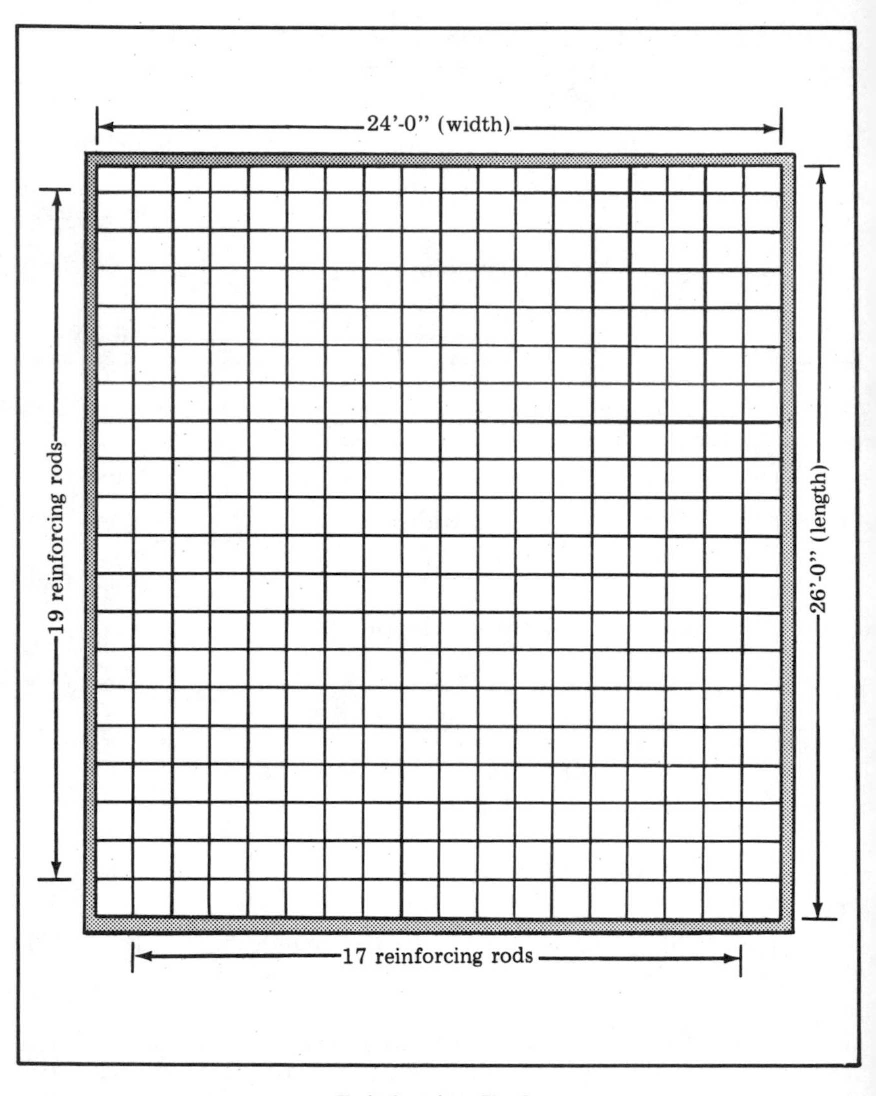

Reinforcing Rods
(16" x 16")
Figure 14-4

Screeds made from 1" x 3" material can act as a thickness and leveling guide. They should be set with a transit or builder's level and held in place with stakes driven firmly into the ground. Allow 2 man-hours per 100 square feet to construct and remove forms and screeds.

Additives for concrete

Some of the additives the estimator may have to include in the concrete estimate are:

1. Calcium chloride . . . accelerates the setting of concrete

Rod Size		Weight
Diameter Inches	Rod Number	Pounds Per Foot
1/4"	2	.167
3/8"	3	.376
1/2"	4	.668
5/8"	5	1.043
3/4"	6	1.502
7/8"	7	2.044
1"	8	2.670

Note: If the reinforcing rods are sold by weight, compute the total weight and cost as follows:

1. Multiply the total number of linear feet by the weight per foot. *Example:* The weight of 940 lin. ft. of ½" diameter rods is 627.92 lbs. (940 x .668 = 627.92 lbs.).

2. Total weight (rounded off) multiplied by the rate = cost.

Reinforcing Rods
Figure 14-5

2. Air-Entraining Agents . . . improves the workability and durability of concrete and results in increased resistance to frost action

3. Coloring agents

Concrete poured in the winter in areas subject to freezing presents a special problem. Concrete that is allowed to freeze soon after placement gains very little strength and some permanent damage occurs. The result of frost damage is not always visible. Surface scaling is easily recognizable damage. Scaling an inch or more can occur and may become progressive. Frost action also can produce cracking, crumbling or "powdering."

Frozen ground should always be thoroughly thawed before concrete is poured. Two percent calcium chloride can be used in combination with 5% or 6% air-entrainment as additives in the concrete for cold weather protection. After pouring, protect the concrete from freezing with a cover of straw or styrofoam slabs. Temporary heaters used with polyethylene may be necessary to protect the concrete. Good ventilation is important when using temporary heaters to cure floor slabs. Fuel-burning heaters produce carbon dioxide which combines with calcium hydroxide in fresh concrete to form a thin, poorly bonded layer of calcium carbonate on the surface. When this happens, the final floor surface will dust under traffic.

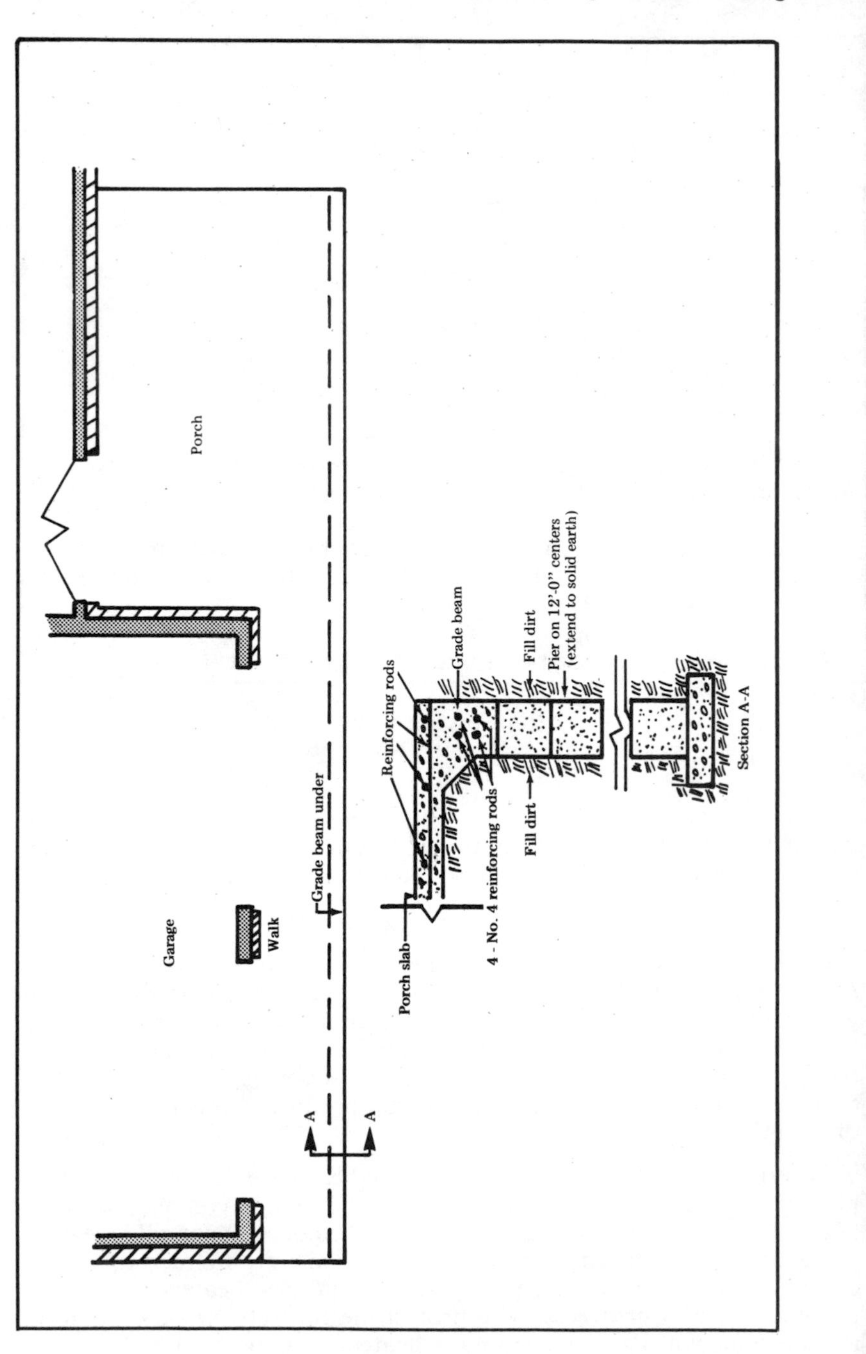

Grade Beam
Figure 14-6

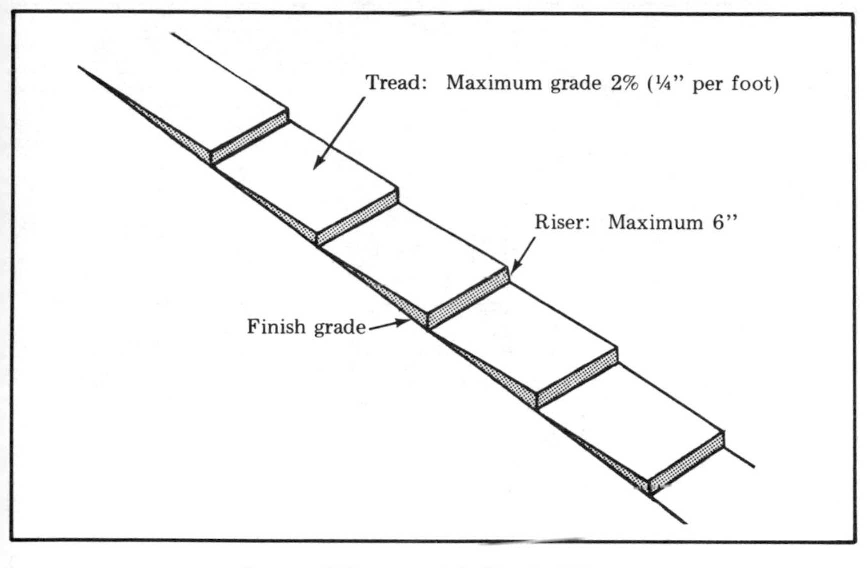

Stepped Ramp with Single Risers
Figure 14-7

When the temperature is below 30°F, concrete is very stiff and difficult to finish. Delay exterior work such as sidewalks, driveways, porches, terraces, patios and garage floors until spring.

Pouring and finishing concrete is specialty work. Those who specialize in it can do the work faster and more cheaply than workmen who do it only occasionally. The subcontractor who specializes in this work normally charges by the square foot. Don't overlook the weather and other conditions that can delay a ready-mix concrete truck from unloading in the time alloted before a penalty charge is added. Cold weather can increase labor costs from 5% to 15%.

The materials for the basement floor, garage and utility room floor, front porch and walk and terrace in Figures 7-1 and 7-2 were estimated as follows:

Basement (928 sq. ft.)

Concrete: (4" thick)
Area times factor for 4" thickness from Figure 14-2 equals cubic yards.
928 sq. ft. x .01235 factor = 11.46 cu. yds.
Order: *12 cu. yds. 3000 psi concrete*

Crushed stone: (4" thick)
Cubic yards times 1.35 factor equals tons.

11.46 cu. yds. x 1.35 = 15.47 tons*

*Note: This crushed stone was placed after the footings were poured and before the foundation walls were started, to save labor later.
Order: *None*

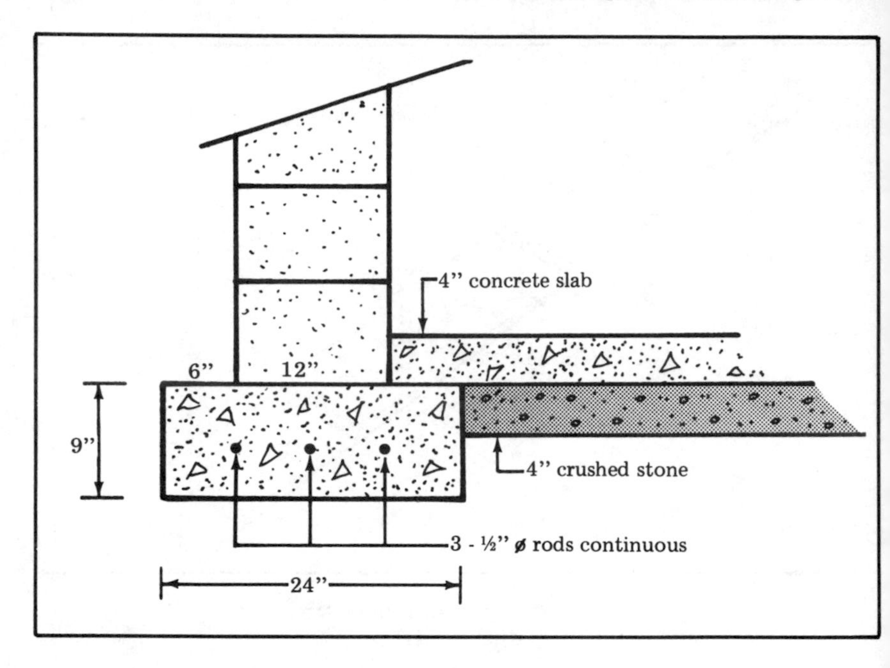

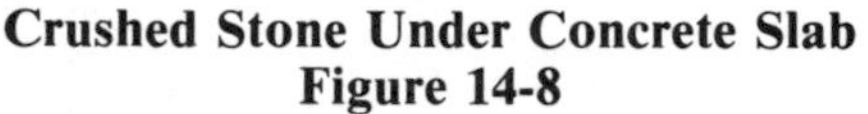

Crushed Stone Under Concrete Slab
Figure 14-8

Wire mesh: (6 x 6 : 10/10

$$\frac{928 \text{ sq. ft.}}{750 \text{ ft. per roll}} = 1.24 \text{ or } 2 \text{ rolls}$$

Order: *2 rolls 6 x 6 : 10/10 wire mesh*

Expansion joints: (½" x 4" x 10')

38'-8" + 24'-0" x 2 = 125'-4"

Order: *130 lineal feet ½" x 4" x 10' expansion felt.*

*Garage and Utility Areas (670 sq. ft.)**

*Note: This area is poured over a fill and #4 reinforcing rods supported by the foundation walls will be used.

Concrete: (4" thick)

670 sq. ft. x .01235 factor = 8.27 cu. yds.

Order: *8.5 cu. yds. 3000 psi concrete*

Crushed stone: None

*Reinforcing rods: (#4 tied @ 16" x 16")**

Order: *1000 lineal feet #4 reinforcing rods (50 pcs.)*

Tie wire:

Allow 1 lb. per 400 lineal feet rods.

$$\frac{1000 \text{ lin. ft.}}{400} \times 1 = 2.50 \text{ lbs.}$$

Order: *2.5 lbs. tie wire*

Expansion joints: (½" x 4" x 10' fiber)

22'-4" + 30'-0" x 2 = 104'-8"

Order: *110 lin. ft. ½" x 4" x 10' fiber expansion joints.*

*Front Porch and Walk (216 sq. ft.)**

*Note: This area is poured over a fill. A grade beam as shown in Figure 14-6 will be required. Number 4 reinforcing rods supported by the house foundation wall and grade beam will be used.

Concrete:

Porch and walk: (4" thick and 216 square feet)

216 sq. ft. x .01235 factor = 2.67 cu. yds.

Grade beam: (8" x 12" x 40'-0")

.67' (8") x 1.0' (12") x 40.0'

Grade beam: (8" x 12" x 40'-0")

$$\frac{.67' (8") \times 1.0' (12") \times 40.0'}{27} = .99 \text{ cu. yds.}$$

Total 3.66 cu. yds

Order: *4 cubic yards 3000 psi concrete*

Crushed stone: None

Reinforcing rods: (#4 tied @ 16" x 16")

Order: *460 lineal feet* #4 reinforcing rods

Tie wire:

Allow 1 lb. per 400 lineal feet rods.

$$\frac{460 \text{ lin. ft.}}{400} \times 1 = 1.15 \text{ lbs.}$$

Order: *1.5 lbs. tie wire*

Terrace (200 square feet)

Concrete: (4" thick)

200 sq. ft. x .01235 factor = 2.47 cu. yds.

Order: *2.5 cubic yards 3000 psi concrete*

Crushed stone: None

Wire mesh (6 x 6 : 10/10)

200 square feet required *

*Note: Enough wire mesh was left over from the basement to use on the terrace.

Order: *None*

Forms and screeds

All material for the forms and screeds, including stakes and nails, will be used from extra material on the job. Therefore no extra material will be required.

Vapor Barrier
Order: *None*

Additives For Concrete
Order: *None*

Winter Protection
Order: *None*

The consolidated list of materials to pour the basement floor, garage and utility room floor, front porch and walk (including the grade beam) and terrace in Figures 7-1 and 7-2 is:

Concrete (3000 psi)

Basement	12.0 cu. yds.
Garage and utility area	8.5 cu. yds.
Front porch and walk	4.0 cu. yds.
Terrace	2.5 cu. yds.
Total	*27.0 cu. yds.*

Crushed Stone

None required

Wire Mesh

Basement and terrace 2 rolls *(6 x 6 : 10/10)*

Reinforcing Rods

Garage and utility area	1000 lin. ft.
Front porch and walk	460 lin. ft.
Total	*1460 lin. ft. (No. 4)*

Tie Wire

Garage and utility area	2.50 lbs.
Front porch and walk	1.50 lbs.
Total	*4.00 lbs.*

Expansion Joints (½" x 4" x 10' fiber)

Basement	130 lin. ft.
Garage and utility area	110 lin. ft.
Total	*240 lin. ft.*

(Enter all of the above listed materials on the Cost Estimate Worksheet)

The carpentry labor to construct and remove the forms and screeds and place the expansion joints for the above work is estimated at 40 man-hours. One carpenter at $8.50 per hour and one laborer at $5.00 per hour will be assigned for this work. The man-hours per workman will be:

$$\frac{\text{40 man-hours}}{\text{2 workmen}} = \text{20 man-hours each}$$

20 hours @ $8.50 =	$170.00
20 hours @ 5.00 =	100.00
Total	*$270.00*

The labor to pour and finish the concrete for the above work was let on contract for $400.00. The subcontractor carries insurance on his employees.

(Enter the above costs on the Cost Estimate Worksheet)

Cost Estimate Worksheet For Concrete Floors, Walks and Terrace

Concrete:

_________ cubic yards @ _________ = $_________

(Test: _________ psi)

_________ cubic yards @ _________ = _________

(Test: _________ psi)

Crushed stone:

_________ cubic yards @ _________ = _________

(Size: _________)

Wire mesh:

_________ rolls/sq. ft. @ _________ = _________

(Size: _________)

Reinforcing rods:

_________ lineal feet @ _________ = _________

(Size: _________)

Tie wire:

_________ lbs. @ _________ = _________

Expansion joints:

_________lineal feet @ _________ = _________

(Type: _________ ; Size: _________)

Vapor barrier:

_________ rolls @ _________ = _________

(Size: _________)

Additives for concrete _________

(List on separate sheet and enter total cost here)

Forms and screeds _________

(List on separate sheet and enter total cost here)

Winter protection _________

(List any materials needed or cost of temporary heat to protect the concrete from freezing, and enter total cost here)

Other material ________

(List on separate sheet and enter total cost here)

Subtotal $________ $________

Sales tax (____ %) ________

Cost of materials $________

Labor:

Carpentry $________

Subcontract ________

Total $________ ________

Cost of labor and materials $

(Enter on Line 14, Form 100)

Chapter 15
Interior Trim

After the drywall or plaster has been installed and the walls have dried, the interior trim work can begin. The flooring and floor underlayment must be installed before the interior doors and baseboard. The interior trim in this chapter will include the following:

1. Flooring
2. Floor underlayment
3. Interior doors, complete with trim and hardware
4. Window trim
5. Baseboard
6. Base shoe
7. Wall molding
8. Paneling and molding
9. Kitchen cabinets and counter tops
10. Vanities, bars and tops
11. Closet shelves
12. Stairs
13. Mirrors and medicine cabinets
14. Tub and shower doors

15. Bathroom accessories

16. Miscellaneous materials

Flooring

Wood flooring applied over a subfloor is manufactured from both hardwood and softwood trees. Generally broad leaves are characteristic of hardwood trees and needles of softwood trees. This does not indicate the degree of hardness of the wood. Oak and maple are the most popular hardwood flooring. Southern pine and Douglas fir are the most popular softwood flooring.

Flooring comes in three types: strip, plank and blocks. Strip flooring is available in widths of 1½" and 2", in thicknesses of ⅜" and ½". For 25/32" thickness the widths are 1½", 2", 2¼" and 3¼". The length of the flooring varies from 1'0" to 16'0". The most commonly used strip flooring is 25/32" x 2¼" tongue and grooved-end matched. The weight of 25/32" x 2¼" oak or maple tongue and grooved-end matched flooring is approximately 2100 lbs. per thousand square feet.

To estimate 25/32" x 2¼" tongue and grooved-end matched flooring, multiply the area to be covered in square feet by the factor of 1.38. This is the number of board feet of flooring needed. This factor allows for side and end matching and 5% for waste. Figure 15-1 shows the square feet for different room sizes.

Flooring should be nailed from 10" to 12" o.c. with 7d (2¼") flooring nails, or machine nailed with 2" power cleats. Allow 20 lbs. 7d spiral flooring nails, or 1 box (5000) 2" power cleats per 1000 square feet for 25/32" x 2¼" flooring.

Plank flooring will most likely be random width and have wooden plugs of a contrasting color. This imitates the flooring used in colonial times.

Block or parquet flooring is made of pieces of strip flooring bonded together edgewise to form a square unit, usually 9" x 9". They usually have tongue and groove edges and can be either nailed to the subfloor or laid in mastic over concrete or wood.

Wood flooring can either be unfinished or have a factory finish. If the flooring is unfinished it must be finished by sanding and applying a filler, stain and finish. Floor finishing should be done by specialists in this work. They will contract for labor and material at a designated price per square foot.

The labor to install flooring will vary from 4 man-hours per 100 square feet for 25/32" x 2¼" flooring to approximately 10 man-hours for parquet flooring.

Floor underlayment

If the finish floor is to be carpet, tile or sheet vinyl, a floor underlayment should be installed over the subfloor and under the floor covering. This underlayment can be particleboard or plywood. On floors in kitchens, bathrooms and utility areas where excess moisture may be a problem, use

Length in Feet

Width in Feet

	2'	3'	4'	5'	6'	7'	8'	9'	10'	11'	12'	13'	14'	15'	16'	17'	18'	19'	20'
2'	4	6	8	10	12	14	16	18	20	22	24	26	28	30	32	34	36	38	40
3'	6	9	12	15	18	21	24	27	30	33	36	39	42	45	48	51	54	57	60
4'	8	12	16	20	24	28	32	36	40	44	48	52	56	60	64	68	72	76	80
5'	10	15	20	25	30	35	40	45	50	55	60	65	70	75	80	85	90	95	100
6'	12	18	24	30	36	42	48	54	60	66	72	78	84	90	96	102	108	114	120
7'	14	21	28	35	42	49	56	63	70	77	84	91	98	105	112	119	126	133	140
8'	16	24	32	40	48	56	64	72	80	88	96	104	112	120	128	136	144	152	160
9'	18	27	36	45	54	63	72	81	90	99	108	117	126	135	144	153	162	171	180
10'	20	30	40	50	60	70	80	90	100	110	120	130	140	150	160	170	180	190	200
11'	22	33	44	55	66	77	88	99	110	121	132	143	154	165	176	187	198	209	220
12'	24	36	48	60	72	84	96	108	120	132	144	156	168	180	192	204	216	228	240
13'	26	39	52	65	78	91	104	117	130	143	156	169	182	195	208	221	234	247	260
14'	28	42	56	70	84	98	112	126	140	154	168	132	196	210	224	238	252	266	280
15'	30	45	60	75	90	105	120	135	150	165	180	195	210	225	240	255	270	285	300
16'	32	48	64	80	96	112	128	144	160	176	192	208	224	240	256	272	288	304	320
17'	34	51	68	85	102	119	136	153	170	187	204	221	238	255	272	289	306	323	340
18'	36	54	72	90	108	126	144	162	180	198	216	234	252	270	288	306	324	342	360
19'	38	57	76	95	114	133	152	171	190	209	228	247	266	285	304	323	342	361	380
20'	40	60	80	100	120	140	160	180	200	220	240	260	280	300	320	340	360	380	400

Square Feet In Room
(Use For Floor And Ceiling)
Figure 15-1

plywood with waterproof glue. The most common size floor underlayment in new construction is ⅝'' x 4' x 8'. To estimate floor underlayment, divide the area in square feet by the coverage of one sheet (32 sq. ft. for a 4' x 8' sheet) and round off to the next whole number. See Figure 15-1 for square feet in different size rooms. *Example:* a kitchen with 120 square feet is to have ⅝'' x 4' x 8' floor underlayment.

$$\frac{120}{32} = 3.75 \text{ or 4 pieces needed}$$

Floor underlayment should be nailed 6'' o.c. on the intermediate supports and 3'' o.c. around the edges. Use 6d (2'') ring grooved nails, and allow 2 lbs. per 100 square feet.

The labor to install the floor underlayment may be estimated at 2 manhours per 100 square feet.

Interior doors

Interior doors are normally hollow-core flush or paneled, and are 1⅜'' thick. The complete door unit consists of the door frame, door, door butts, door stops, casing on both sides and the door lock. Prehung doors are the most popular and are widely used because they reduce labor costs. They come prehung in the door frame with the door butts, door stops and casing installed. The unit may not include the door lock, but the cross hole and bolt hole will be drilled for the standard 2⅜'' backset for a cylindrical lockset.

When estimating interior doors, the number and size can be determined from the floor plan or the door schedule. The specifications will determine the type of doors to use.

Use 8d finish nails for installing the interior door casing. Allow ½ lb. per opening.

The labor to set a pre-hung door, install and sand the joints and install the door lock may be estimated at 3 man-hours per door. For other type interior doors such as sliding or bi-fold, estimate 4 to 5 man-hours per door to assemble and set the door frame, install the casing on two sides and install the hardware and door.

Window trim

The side and head casing of windows are normally the same design and kind of wood as the door trim. The window trim consists of casing, stool, apron, stops, window lock and window lift. Some windows are designed so the stops can be used on four sides, thus eliminating the stool. The casing is placed on all four sides, giving the window a picture frame appearance. Metal windows do not normally require casing, only a window sill. On these windows the drywall or plaster makes a return to the metal window, thus eliminating the casing and apron.

Window trim can be purchased by the linear foot, or in sets for each window. If the trim is purchased in linear feet, the estimator must add an allowance of 20% for cutting and waste. Many manufacturers of wood windows also manufacture window trim. The trim will probably be sold in sets for each size window. The set will include the casing, stops, mullions, stool and apron (if required). No allowance for waste is necessary when the window trim is purchased in sets. The number and type of windows, the trim, locks and window lifts are shown in the plans, the window schedule and specifications. Figures 15-2 and 15-3 show some common window and door trim details.

The finishing nails for the window trim should be 6d (2'') or 8d (2½''). All joints should be glued, nailed and sanded before they are finished. Allow ¼ lb. of nails for each side.

The labor to install the window trim will vary, depending on the size and type of window. A 3' x 3' single window unit will require about 50% less labor to trim than a large triple window or picture window. Previous cost records are your best guide. Generally,you can allow 3 man-hours per window, regardless of size.

Baseboard and Baseshoe

The baseboard runs continuously around the perimeter of each room, hall and closet and hides the space between the finish floor and the plaster or wallboard. It is not necessary where there is wood paneling. Baseboard is made of the same wood as the window and door trim. All joints in baseboard should be over a stud to provide a firm nailing base. Add an allowance of 10% for cutting and waste.

Baseboard is sold in linear feet. Total the linear feet of each room, hall and closet. Figure 15-4 shows the perimeter of rooms in linear feet. Door openings are not normally deducted.

Baseshoe is used with wood flooring and resilient flooring made of asphalt, vinyl asbestos, vinyl, rubber and linoleum, either in sheets or tile. It is also used around the perimeter of paneled walls with or without baseboard. Where wall-to-wall carpet will be used, it is not needed. Estimate it the same way as baseboard. Figure 15-7 shows some types and sizes of baseboards and baseshoe.

Use 8d (2½'') finish nails for the baseboard and allow 1 lb. per 100 linear feet. For the baseshoe, use 6d (2'') finish nails and allow ½ lb. per 100 linear feet.

Estimate 4 man-hours per 100 linear feet for installation of the baseboard, including finishing the joints. For the baseshoe, estimate 2 man-hours per 100 linear feet.

Wall molding

One built-in aesthetic feature the builder can add to give a room a luxurious look is wall molding feature such as chair rails and panel mold. These can be applied to the walls of any room in the house, but are usually used in living rooms, dining rooms, family rooms and foyers.

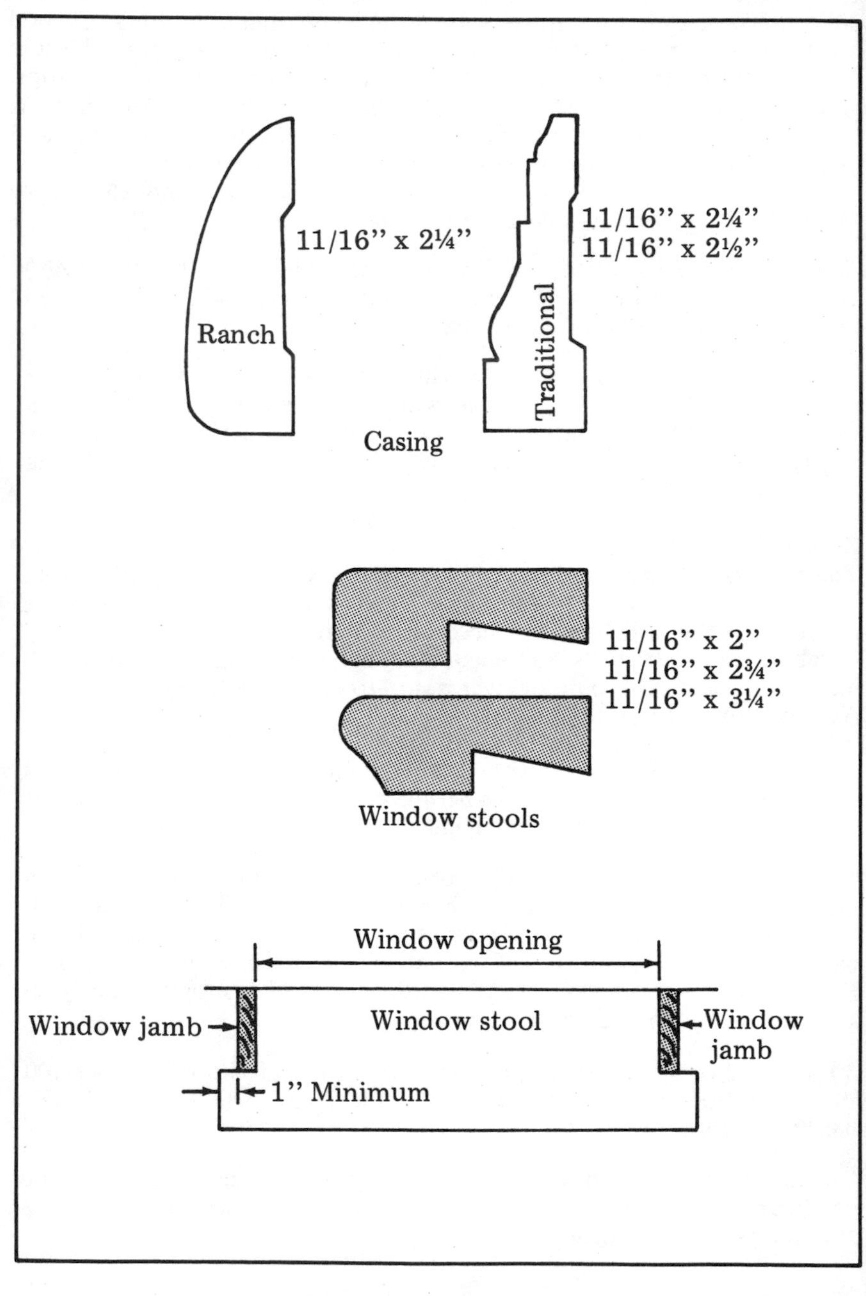

Window and Door Trim
Figure 15-2

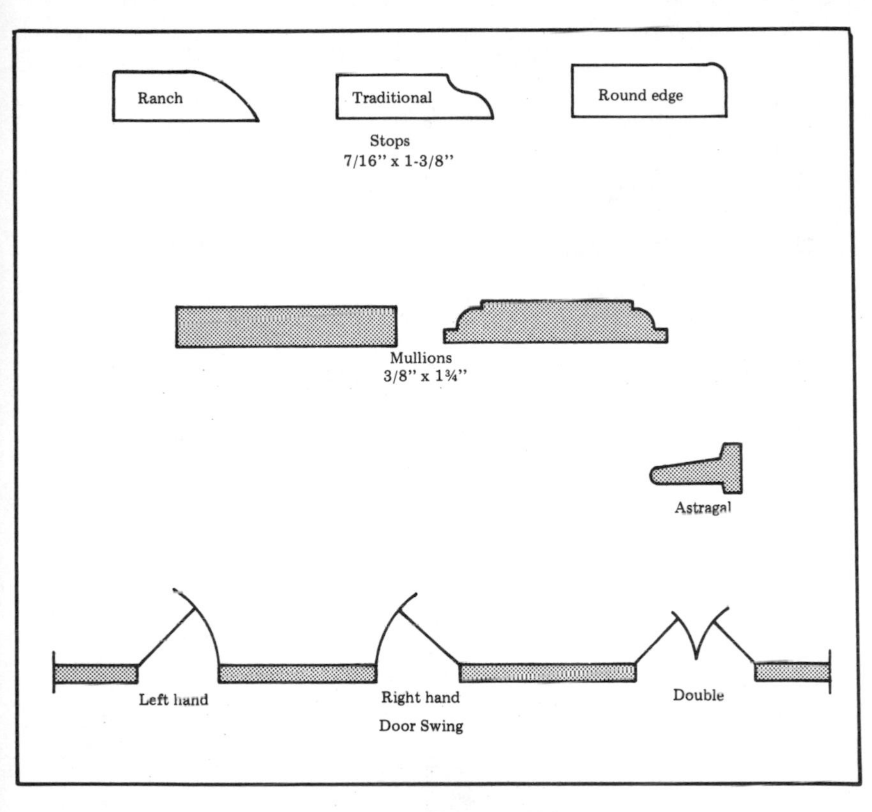

Window and Door Trim
Figure 15-3

Figure 15-5 shows ways to install this molding.

1. Chair rail (Figure 15-6) can be purchased in a variety of designs and sizes. The recommended height from the floor to the top of the chair rail is 32". A good grade of adhesive should be used together with nails to prevent the chair rail from separating from the wall between the studs where there are no nails. This chair rail can be used on all walls in a room or on just one. A mural or other scenic wall fabric on the wall above the chair rail can be very effective.

2. Panel mold is applied to the wall between the chair rail and baseboard. It should also be applied with a good grade of adhesive together with nails. The length and height of the designs that can be fashioned from this mold are flexible. However, there are some guidelines to follow to give the design a professional look. As shown in Figure 15-5, the distance from the inside corner to the panel mold (a) and (b) should be equal. This distance may vary from 4" to 6". If the chair butts into a window or door casing between the two opposite walls, the distance from this casing to the panel mold should also be equal to the

Length in Feet

Width in Feet

	2'	3'	4'	5'	6'	7'	8'	9'	10'	11'	12'	13'	14'	15'	16'	17'	18'	19'	20'
2'	8	10	12	14	16	18	20	22	24	26	28	30	32	34	36	38	40	42	44
3'	10	12	14	16	18	20	22	24	26	28	30	32	34	36	38	40	42	44	46
4'	12	14	16	18	20	22	24	26	28	30	32	34	36	38	40	42	44	46	48
5'	14	16	18	20	22	24	26	28	30	32	34	36	38	40	42	44	46	48	50
6'	16	18	20	22	24	26	28	30	32	34	36	38	40	42	44	46	48	50	52
7'	18	20	22	24	26	28	30	32	34	36	38	40	42	44	46	48	50	52	54
8'	20	22	24	26	28	30	32	34	36	38	40	42	44	46	48	50	52	54	56
9'	22	24	26	28	30	32	34	36	38	40	42	44	46	48	50	52	54	56	58
10'	24	26	28	30	32	34	36	38	40	42	44	46	48	50	52	54	56	58	60
11'	26	28	30	32	34	36	38	40	42	44	46	48	50	52	54	56	58	60	62
12'	28	30	32	34	36	38	40	42	44	46	48	50	52	54	56	58	60	62	64
13'	30	32	34	36	38	40	42	44	46	48	50	52	54	56	58	60	62	64	66
14'	32	34	36	38	40	42	44	46	48	50	52	54	56	58	60	62	64	66	68
15'	34	36	38	40	42	44	46	48	50	52	54	56	58	60	62	64	66	68	70
16'	36	38	40	42	44	46	48	50	52	54	56	58	60	62	64	66	68	70	72
17'	38	40	42	44	46	48	50	52	54	56	58	60	62	64	66	68	70	72	74
18'	40	42	44	46	48	50	52	54	56	58	60	62	64	66	68	70	72	74	76
19'	42	44	46	48	50	52	54	56	58	60	62	64	66	68	70	72	74	76	78
20'	44	46	48	50	52	54	56	58	60	62	64	66	68	70	72	74	76	78	80

*Note: No door openings are deducted

Perimeter of Room in Lineal Feet *
(Use For Base And Shoe)
Figure 15-4

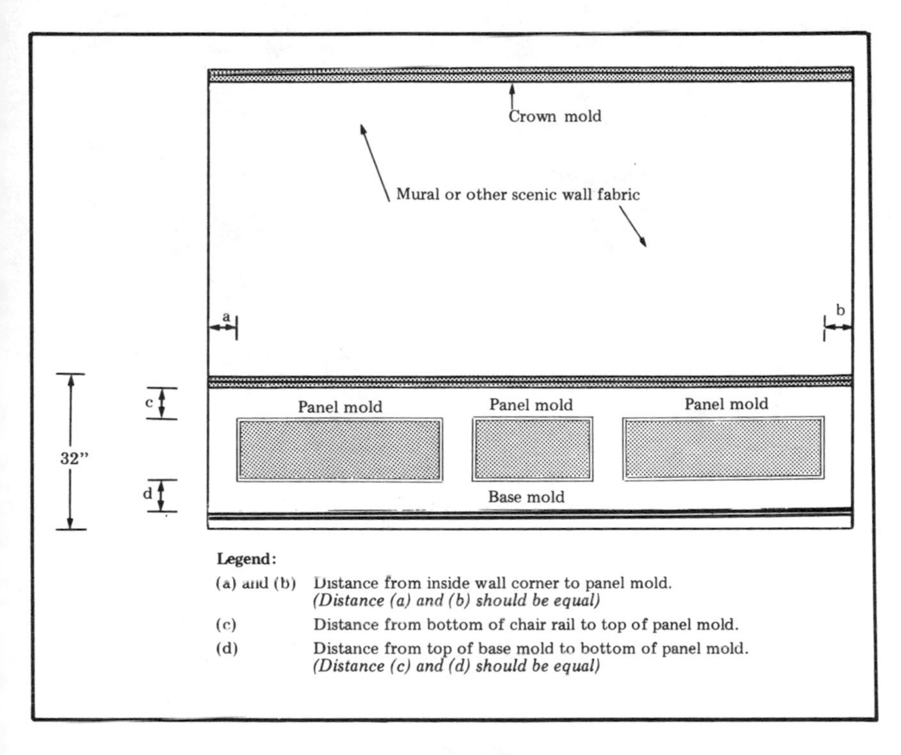

Wall Molding
Figure 15-5

opposite end as shown at (a) and (b). The distance from the bottom of the chair rail to the top of the panel mold (c) should be equal to (d), the distance from the top of the base mold to the bottom of the panel mold. This distance should be the same as (a) and (b) for a perfectly balanced design. After the panel mold has been installed, the wall can be decorated with paint or wall fabric.

Chair rail and panel mold is estimated by the linear foot. Find the linear feet of the walls where the chair rail is to be used and the linear feet of design patterns for the panel mold. Add 10% for waste.

Use 8d (2½'') finish nails with adhesive for the chair rail and allow 1 lb. nails per 100 linear feet. Use 6d (2'') finish nails with adhesive for the panel mold and allow ½ lb. nails per 100 linear feet.

Allow 6 man-hours per 100 linear feet of chair rail and panel mold for the labor.

Paneling
A wood paneled wall imparts a sense of warmth and beauty. It is decorative and increases the strength of a wall. Hardboard,plywood and solid wood are the three basic types of paneling.

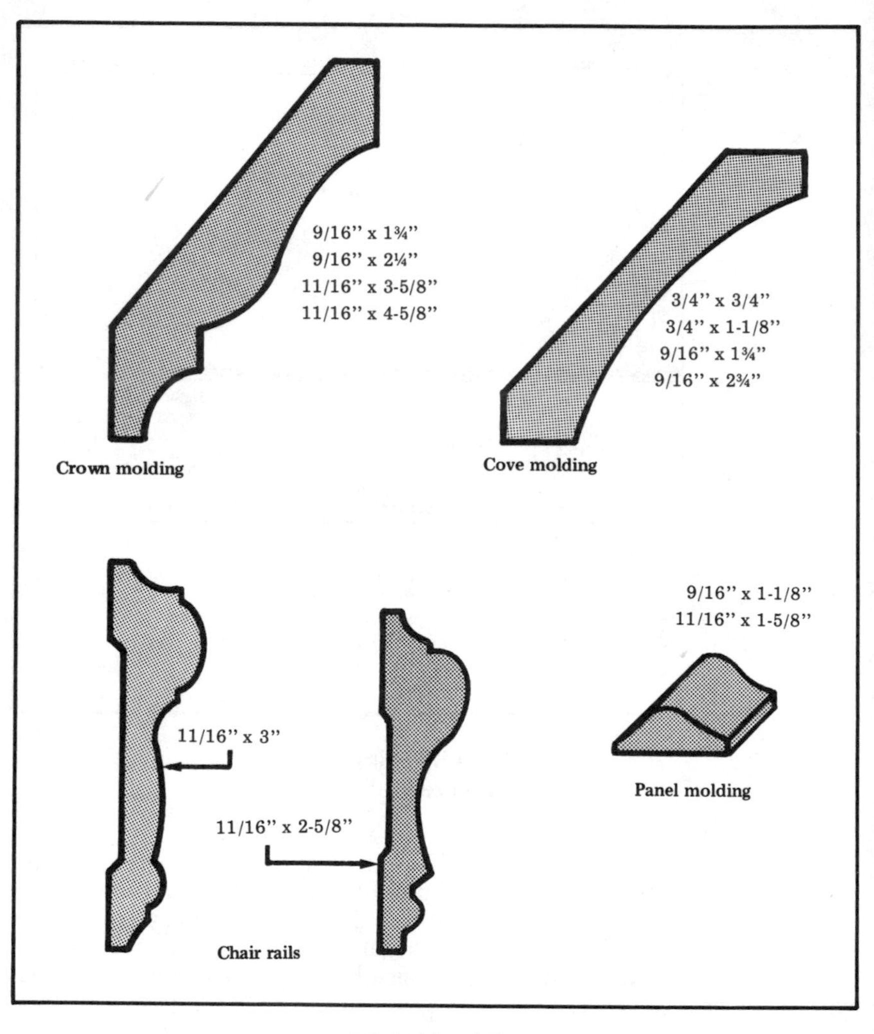

Wall Moulding
Figure 15-6

Hardboard paneling is made from wood fibers. It is smooth, hard and dense. It is available in ⅛" or ¼" thickness and in various sizes, the most popular being the 4' x 8', or 4' x 10' size. The surface of hardboard paneling is usually finished in a simulated wood-grain to imitate various woods.

Plywood paneling is usually ¼" x 4' x 8' and can have a factory finish and V-grooves at random widths.

Solid wood paneling is made from both hardwood and softwood. It normally has tongue and groove joints and comes in random widths. It is

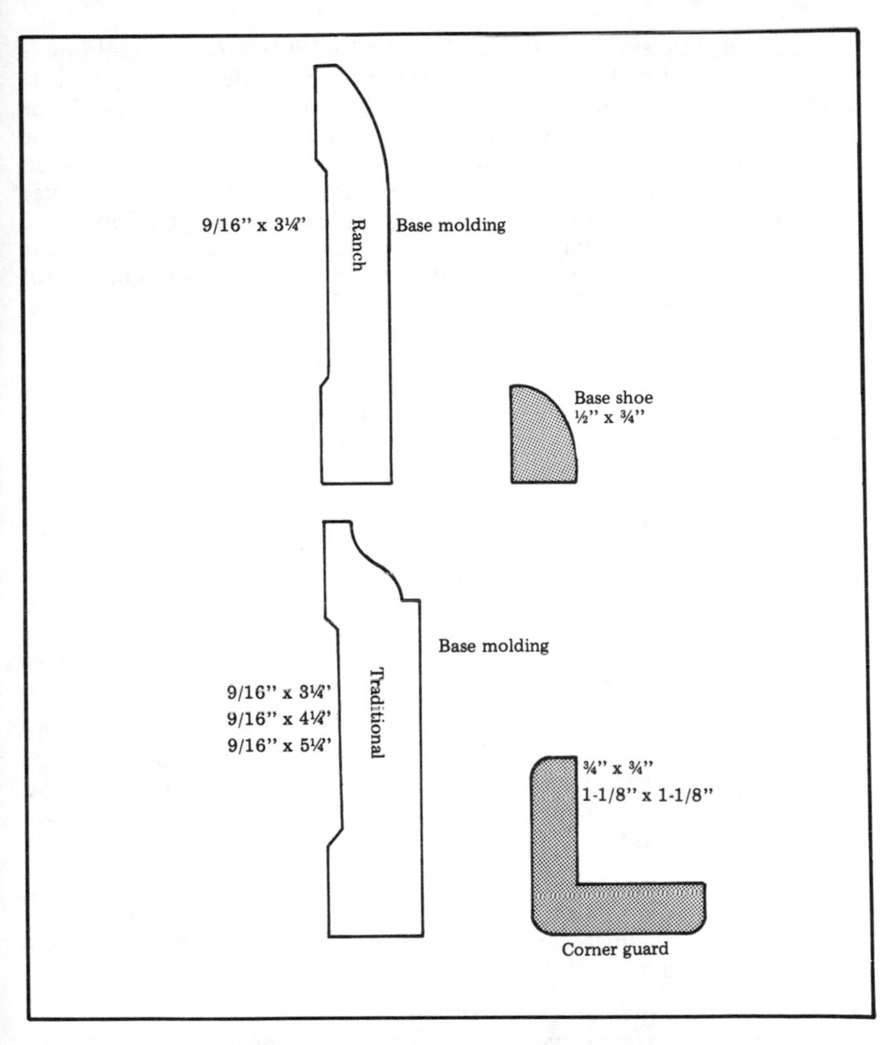

Base Molding and O/S Corner Guard
Figure 15-7

graded either knotty or clear. The knotty wood paneling has a rustic appearance. The clear wood paneling looks more formal.

To estimate hardboard or plywood paneling in 4' widths, divide the perimeter of the room (see chart in Figure 15-4) by 4' and round off to the next whole number. *Example:* the perimeter of a room with dimensions of 12' x 15' has 54 linear feet.

$$\frac{54}{4} = 13.50 \text{ or } 14 \text{ pieces}$$

Inside corners are normally scribed. Special inside corner molding is available prefinished to match the paneling. Outside corners are best finished with prefinished outside corner molding or corner guards, as shown in Figure 15-7. The corner molding is 8' in length. To estimate the number of pieces, count the corners where they will be used. Prefinished molding used at the ceiling is sold by the linear foot. Find the perimeter of the room where molding will be used, divide by the length of one piece and round off to the next whole number. *Example:* if there are 54 linear feet in the perimeter of a room and the molding is available in 10' lengths, the molding will be estimated as follows:

$$\frac{54}{10} = 5.40 \text{ or 6 pieces (60 linear feet)}$$

The nails for ¼" hardboard or plywood paneling can be 1⅝" colored nails (usually packed in ¼ lb. boxes) or 4d (1½") finish nails. Allow 1 box of colored nails or 1 lb. 4d finish nails per 100 square feet. When panel adhesive is used, allow 1 tube of adhesive per 100 square feet. Nail holes should be set and filled with colored putty if colored nails are not used.

When estimating solid wood paneling, add an allowance for the tongue and groove and the waste. Normally the area in square feet multiplied by the factor of 1.25 will indicate the amount to order. Use 6d (2") finish or casing nails. Allow 2 lbs. per 100 square feet.

When the type of paneling has not been selected at the time the estimate is made, make an allowance for the paneling. If later, the owner selects paneling that is higher than the allowance, the difference will be an extra charge. If the paneling costs less than the allowance, the owner will be given a credit.

The labor to install hardboard or plywood paneling may be estimated at 3 man-hours per 100 square feet. Estimate molding installation at 5 man-hours per 100 linear feet. Estimate the labor for solid wood paneling at 6 man-hours per 100 square feet.

Kitchen cabinets and vanities

The floor plan, together with the kitchen and bathroom detail section, will show the linear feet, and the type and size of cabinets required. The specifications indicate the cabinet material and the type of counter tops to use. If other cabinets are to be installed, such as in a wet bar, they will be shown on the floor plans with a detail section. They also will be described in the specifications. Often an allowance will be made for all cabinets in a house because the owner may not know the exact type of cabinets he wants at the time the estimate is prepared.

The labor to install factory-built cabinets, including the counter tops and molding, can be estimated at 7 man-hours per 100 square feet of face area.

Closet shelves

The floor plan of the blueprints shows the size of the closets. For clothes closets the floor plan will indicate the shelf and rod with a dotted line under the shelf for the rod. For linen closets the number of shelves will be indicated.

The material for clothes closets consists of 12'' shelving material, 1' x 4'' cleats or ribbons, clothes rod and supports, and rosettes. The shelf is normally 66'' from the floor. If the material is purchased separately, scale the length of the closet from the floor plan and order the shelves and cleats in linear feet. Purchase the clothes rod for the length of the closet, and allow one pair of rosettes per rod. If the clothes rod is longer than 48'', one or more rod supports will be needed.

The material for linen closets consists of shelving material and cleats. When estimating the material in linear feet, order in lengths that require the least amount of waste. *Example,* a linen closet 3'0'' in length requires 5 shelves. If 1 - 1'' x 12'' x 16' was ordered, the 5 shelves could be cut from this one piece of material with only 1'0'' waste. If the shelving were ordered in 8' lengths, it would require 3 - 1'' x 12'' x 8' pieces, resulting in 9'0'' waste.

Prehung shelves are widely used today. They come complete with all hardware for any type of closets, and are adjustable for different closet lengths. When estimating prehung closet shelves, list each closet separately and order the prehung shelves as required for each closet length.

Estimate the labor for closet shelves at 3 man-hours per 100 linear feet of shelving.

Stairs

The material for the stairs consists of the stringers or carriage, treads, risers, skirtboard or finish stringer, and molding (Figure 15-8). A balustrade consisting of the balusters, handrail, turnout, newel post, and rosettes (Figure 15-9), is used on the open side of the main stairs. If the stair opening is enclosed by a wall on each side, only a handrail and handrail brackets are needed. For basement stairs, risers are optional. The handrail is secured to the wall with handrail brackets at the top of the stairs and a post (normally 4'' x 4'') at the bottom.

The stringers for the main stairs are normally constructed during the framing-in of the house for the safety and convenience of the workmen. Temporary treads are used until they are finished with the interior trim. Three stringers are needed for each set of stairs 30'' or more in width. The length of the stringers varies with the number and dimensions of the risers and treads. Figure 7-38 will help you determine the length of stringers.

When estimating material for stairs, refer to the blueprints for the floor-to-floor rise. If the number of risers is not shown on the plans, it can be calculated by converting the floor-to-floor rise (see Figure 7-39) into inches, dividing by 8 and rounding off. *Example,* in Figure 7-39 the floor-to-floor rise is 8'10⅜''. This is 106⅜'' (106.38'').

$$\frac{106.38''}{8} = 13.30 \text{ (use either 13 or 14 risers)}$$

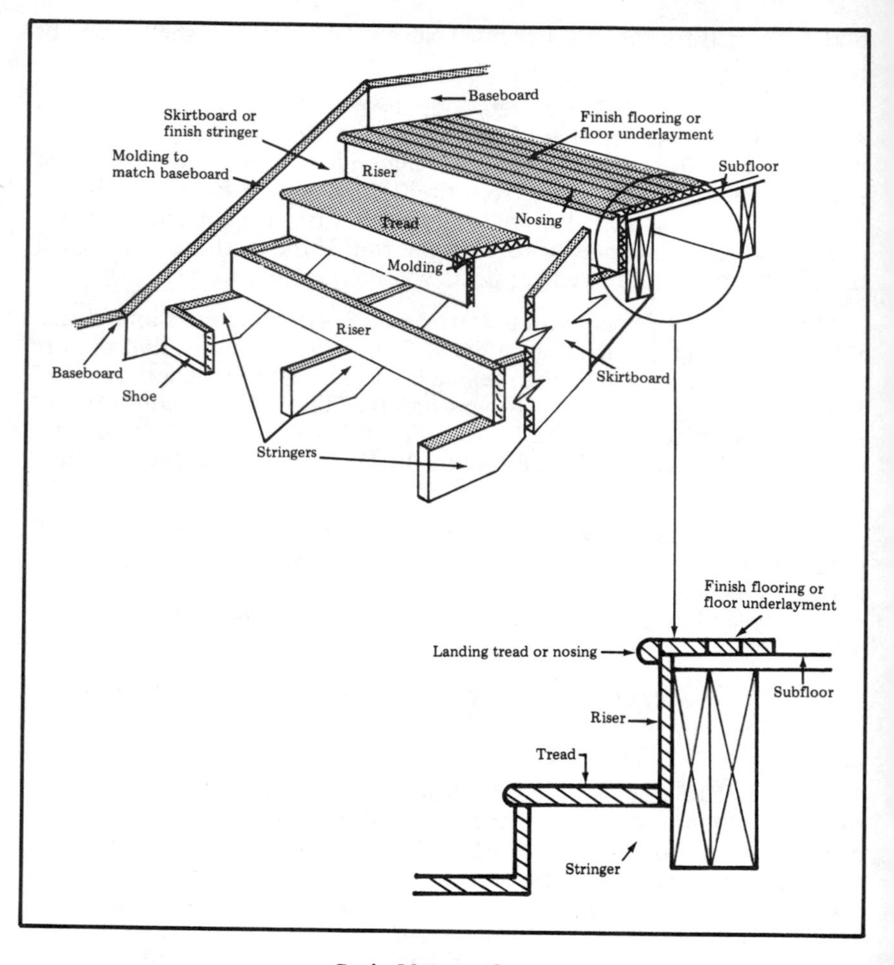

Stair Nomenclature
Figure 15-8

If 13 risers are used, each riser will be 8³⁄₁₆", which is near the maximum permissible riser height of 8¼" for main stairs. If 14 risers are used, each riser will be 7⅝", which is more acceptable. There is always one less tread than riser. If 14 risers are used, there will be 13 treads. *Stair Builders Handbook* published by Craftsman Book Company has complete stair layout tables.

Once the numbers of risers and treads are known, the material for the risers, treads, skirtboards, molding, handrails, handrail brackets, posts for basement stairs and all material for the balustrade can be estimated. If there are to be landings, they will be on the plans. Estimate this material along with the stairs.

The labor to frame and install the risers, treads and handrails will vary with the type of stairs and the efficiency of the craftsmen. Only craft-

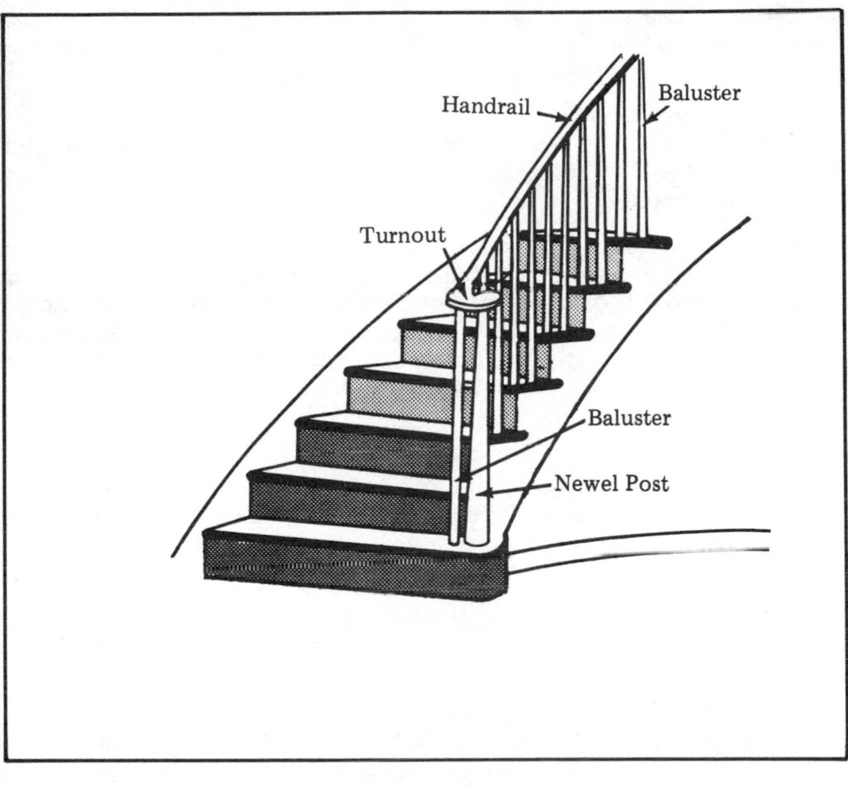

Circular Stair Balustrade
Figure 15-9

smen experienced in stair building should be used on stair work. The labor for basement stairs with open risers and two handrails may be as low as 10 man-hours per set of stairs. Circular stairs and other ornate main stairs with an elaborate balustrade system may required 80 or more man-hours. If you don't have cost records for similar stairs, you'll have to use some judgement in estimating the labor.

Mirrors and Medicine cabinets

Mirrors add glamor to a bathroom. They accent certain architectural features, and can double the apparent size of a small room. They can be used over vanities in bathrooms to replace medicine cabinets or on side walls adjacent to the cabinet.

Mirrors and medicine cabinets will be shown on the floor plan. Detail sections will show the number and size or the specifications will give this information.

The labor to install mirrors and medicine cabinets may be estimated at 8 man-hours per 55 square feet of mirror area. You can use the factor of

0.14545 man-hours per square foot of mirror area. *Example,* the labor to install a mirror 3' x 7' (21 sq. ft.) would be estimated as follows:

$$0.14545 \times 21 = 3.05 \text{ or } 3 \text{ man-hours}$$

Tub and Shower Doors

The number, type and size of tub and shower doors are shown on the floor plans, detail sections and specifications. Allow 2 man-hours per door for installation.

Bathroom accessories

The bathroom accessories include paper holders, towel racks, soap dishes, grab bars, and shower curtain rods (if a shower door is not used). The number, type and style will be selected by the owner. An allowance for the bathroom accessories is normally made. Allow ¼ man-hour per unit for the labor to install the accessories.

Miscellaneous

This will include any special items not shown on the prints which the owner wants. Valances, room dividers, and artificial beams would be included here. Additional materials such as nails, glue, adhesive, wood filler, sandpaper and tub and tile caulking not listed elsewhere may also be listed here.

The materials for the interior trim for the house in Figures 7-1 and 7-2 are estimated as follows:

Flooring

The finish flooring will be wall-to-wall carpet in all areas except the vestibule, kitchen, bathrooms and utility room where vinyl sheet goods will be used. Plywood floor underlayment will be used in the vestibule, kitchen and bathrooms number 1 and 2. The utility area and bathroom number 3 have a concrete floor. Particleboard will be used on the floors where wall-to-wall carpet will be installed. The area where the plywood underlayment is used is 296 square feet.

$$\frac{296}{32} = 9.25 \text{ or } 10 \text{ pieces}$$

Order: *10 pieces 5/8" x 4' x 8' plywood underlayment*

The area where the particleboard is used is 1760 square feet.

$$\frac{1760}{32} = 55 \text{ pieces}$$

Order: *55 pieces 5/8" x 4' x 8' particleboard*

The nails are estimated at 2 lbs. of 6d ring grooved nails per 100 square feet.

$$\frac{2056}{100} \times 2 = 41.12 \text{ or } 42 \text{ lbs.}$$

Order: *42 lb. 6d ring grooved nails*

Interior doors
The interior doors are prehung and include the door frame, stops, door, all hardware including the lock, metal door stops and casing. They are listed as follows:

1 - 1/6 x 6/8 L.H.
2 - 2/0 x 6/8 L.H.
3 - 2/4 x 6/8 R.H.
1 - 2/4 x 6/8 L.H.
6 - 2/6 x 6/8 R.H.
2 - 2/6 x 6/8 L.H.
2 - 2/6 x 6/8 Double
2 - 2/6 x 6/8 Bi-fold
2 - 3/0 x 6/8 Bi-fold
2 - 4/0 x 6/8 Bi-fold
2 - 5/0 x 6/8 Bi-fold

Estimate ½ lb. 8d finish nails per door opening.

Order: *13 lbs. 8d finish nails*

Window trim
The windows except for the picture windows are wood double hung. Window trim is purchased by the set for each window. Each set includes the casing, stops, mullions (where needed), stool, apron, window locks and window lift. The window trim is ordered as follows:

2 - sets 2/0 x 3/2
6 - sets 2/4 x 3/2
1 - set 2/4 x 3/2 Double
1 - set 2/4 x 4/2 Double
1 - set 54-55 Picture window
1 - set for combination picture and double hung window in kitchen.

Estimate ¼ lb. of 8d finish nails per side.

Order: *3 lbs. 8d finish nails*

Baseboard and Baseshoe
Baseboard is used throughout the house except in the family room where paneling is the finish wall. The baseshoe is used only in the areas where the vinyl sheet goods are used. Baseboard and baseshoe are estimated using Figure 15-4.

Order: *850 linear feet baseboard*
250 linear feet baseshoe

Allow 1 lb. of 8d finish nails per 100 linear feet for the baseboard, and ½ lb. of 6d finish nails per 100 linear feet for the baseshoe.

Order: *9 lbs. of 8d finish nails*
2 lbs. of 6d finish nails

Wall molding
None

Paneling
Prefinished plywood paneling ¼" x 4' x 8' is to be used on the walls in the family room. Prefinished cove molding is used at the ceiling and

prefinished outside molding is used at the corners. There is a wall perimeter of 54'6" where the paneling will be installed. The number of panel sheets is estimated as follows:

$$\frac{54.5' \ (54'\text{-}6")}{4} = 13.63 \text{ or } 14 \text{ pieces}$$

Order: *14 pieces ¼" x 4' x 8' prefinished plywood paneling*

The prefinished cove molding at the ceiling will extend around the fireplace wall.

Order: *7 pieces (70 lin. ft.) prefinished cove molding*
3 pieces prefinished o/s corners
5 boxes 1⅝" colored nails
5 tubes panel adhesive

Kitchen cabinets and Vanities
The kitchen cabinets, vanities and tops are furnished per plans and specifications by the manufacturer.

Closet shelves
All closet shelves are prehung and furnished per plans and specifications by the manufacturer.

Stairs
All materials including the stringers, treads, risers, skirtboards, molding, landing tread and the balustrade for the main stairs are furnished by the manufacturer. The material for the basement stairs was not included in the house package. The basement stairs will be estimated as follows:

Floor to floor height . . .8'5½"
14 risers @ 7¼"
13 treads @ 10"

Order: *3 - 2" x 12" x 14' (stringers)*
13 - Fir treads @ 3'6"
14 - Fir risers @ 3'6"
1 - 4" x 4" x 8' cedar post (makes 2 - 4' posts)
2 pieces 1½" x 12' handrail
3 pairs handrail brackets

Mirrors and Medicine cabinets
The plans and specifications indicate a 4' x 3' mirror installed over the vanity in bathroom #1 and a 7' x 3' mirror over the vanity in bathroom #2. No medicine cabinets are specified in these two bathrooms. A medicine cabinet is specified in bathroom #3 with no separate mirror.

Order: *1 - 4' x 3' plate glass mirror*
1 - 7' x 3' plate glass mirror
1 - medicine cabinet per plans

Tub and Shower doors
The tubs in bathrooms #1 and #2 and the shower in bathroom #3 will

have shower doors. These doors were included in the price of the plumbing fixtures but will be installed by carpenters.

Order: *2 - 5' shower doors*
1 - 3' shower door

Bathroom accessories
An allowance of $100.00 is made for all bathroom accessories.

Miscellaneous
There are no valances, room dividers, artificial beams or other items not shown on the blueprints. For other miscellaneous materials estimate the following:

Order: *8 lbs. 8d finish nails*
5 lbs. 6d finish nails
2 lbs. 4d finish nails
1 qt. wood glue
15 sheets sandpaper
4 tubes tub and tile caulking

(Enter all of the above materials on the Cost Estimate Worksheet)

Estimate the labor for the interior trim from past cost records for similar work. The labor for interior trim, including trim for the basement door, all stairs, installing the tub and shower doors and bathroom accessories, is estimated at 482 man-hours. One carpenter at $8.50 per hour, two carpenter helpers at $6.25 per hour and one laborer at $5.00 per hour will be assigned to the interior trim. The man-hours per workman will be:

$$\frac{482}{4} = 120.50 \text{ or } 121 \text{ man-hours each}$$

121 hrs. @ $8.50 x 1 =	$1028.50
121 hrs. @ 6.25 x 2 =	1512.50
121 hrs. @ 5.00 x 1 =	605.00
Estimated labor cost for interior trim	$3146.00

(Enter this labor cost on the Cost Estimate Worksheet)

Cost Estimate Worksheet For Interior Trim

Flooring:

_________ sq. ft. @ _________ = $_________

(Grade _________ ; Species _________ ; Size _________)

Floor underlayment:

_________ pcs./sq. ft.)@ _________ = _________

(Material _________ ; Size _________)

_________ pcs./sq. ft. @ _________ = _________

(Material _________ ; Size _________)

Interior doors: _________

(List all doors complete with trim, all hardware and locks and door stops on separate sheet and enter total cost here)

Window trim: _________

(List all window trim including locks and window lifts on separate sheet and enter total cost here)

Baseboard:

_________ lin. ft. @ _________ = _________

(Type _________ ; material _________ ; size _________)

Baseshoe:

_________ lin. ft @ _________ = _________

(Type _________ ; size _________)

Wall molding: _________

(List on separate sheet and enter total cost here)

Paneling:

_________ sq.ft./bd. ft. @ _________ = $_________

(Type _________ ; size _________)

_________ pcs./sq.ft. @ _________ = _________

(Type _________ ; size _________) Note: list molding, nails, adhesive, etc. under miscellaneous.

Kitchen cabinets and tops: _________

(List on separate sheet and enter total cost here)

Vanities and tops: _________

(List on separate sheet and enter total cost here)

Other cabinets: __________

(List on separate sheet and enter total cost here)

Closet shelves: __________

(List on separate sheet and enter total cost here)

Stairs, main: __________

(List all materials on separate sheet and enter total cost here)

Stairs, basement: __________

(List all materials on separate sheet and enter total cost here)

Mirrors:

__________ number @ __________ = __________

(Size __________)

__________ number @ __________ = __________

(Size __________)

Medicine cabinets:

__________ number @ __________ = $__________

(Size __________)

Shower doors:

__________ number @ __________ = __________

(Size __________)

__________ number @ __________ = __________

Bathroom accessories: Allow = __________

Nails:

__________ lbs. @ __________ = __________

(Size __________)

__________ lbs. @ __________ = __________

(Size __________)

__________ lbs. @ __________ = __________

(Size __________)

__________ lbs. @ __________ = __________

(Size __________)

Miscellaneous: __________

(List on separate sheet and enter total cost here)

Subtotal (carry forward to next page) $__________ $ __________

Brought forward from preceding page $ ________

Sales tax (____ %) ________

Cost of materials $ ________

Labor: ________

Cost of labor and material $ ________

(Enter on line 15, Form 100)

Chapter 16

Painting, Floor Covering and Appliances

Painting

Painting provides a decorative coating which will resist weather and protect from damage. The paint or coating selected must be designed for the use proposed. The printed instructions on each can of paint identify the type of exposure (exterior or interior), the type of surface to be covered (wood, metal, masonry, concrete, plaster) and the type of wear the paint is designed for.

Exterior wood surfaces should receive a prime coat when the trim is installed and a minimum of 4.0 mil (.004 inch) finish coat. Normally this requires two finish coats. The top and bottom of exterior wood doors and all wood windows should receive two coats of paint or sealer. The paint used above masonry surfaces should be the low chalking type to avoid streaking and stains.

Interior plaster or gypsum wallboard may be painted or papered. Paint is the easiest and cheapest way to decorate interior walls. Latex paint in flat, semigloss or gloss is popular. It can be applied with brush, roller or spray gun, dries fast, has little odor, and can be cleaned up with water.

For gypsum and plaster walls, one coat sealer plus a minimum of one finish coat are required. The finish coat in kitchens and bathrooms should have a semigloss or gloss finish to provide a durable and washable surface.

Interior wood surfaces should have paint or natural finish. A fill material is usually applied to keep the wood grain from rising in depressions. It should have three coats including the prime coat and be sanded between coats to provide a smooth surface. The top and bottom of all interior doors should receive one coat of paint or sealer.

Wallpaper or wall fabrics may be used throughout the house or only on designated walls. Wallcovers are available in a variety of patterns and materials. Kitchens and bathrooms should have waterproof wallcovers.

Estimate the cost of painting by the size and shape of the building, the number of coats, the type of paint, the number of windows and doors, the wall preparation work, and the sanding and caulking required. Painting is normally done by paint contractors. They will contract for both the labor and materials or the labor only. Cost is estimated by the square foot. It will vary widely from job to job. Most paint contractors prefer to see the house before submitting a bid. But you can get a preliminary bid at the time the estimate is prepared. This preliminary bid may or may not be binding when the house is ready for painting.

Before a contract is let for painting, a clear set of specifications should be written and signed by all parties. These specifications should specify the type of paint to be used, the number of coats, the caulking and sanding, and all other necessary preparation. An inferior paint job can ruin the appearance of an otherwise attractive house.

The wallpaper or wall fabric is always selected by the owner. Seldom will he or she have decided on the design, type and cost of the material at the time the original estimate is prepared. For this reason an allowance is normally made for the labor and material for all wallpaper.

Floor covering

Wall-to-wall carpet is the first choice in floor cover for most residential construction today. The most popular materials are nylon and polyester fibers. Carpet and pad can be laid directly over a plywood subfloor, eliminating the need for any other floor material.

The *rug pad or cushion* is made of animal hair, fibers, rubberized fibers or cellular rubber and bonded to the underside of the carpet or laid separately. Bonded backing is normally used when carpet is installed over concrete floors. On wood floors it is more common to install the backing separately. Cushioning increases the resilience and durability of carpet. It is graded by ounces per square yard — the heavier the cushion, the better it will be.

Carpet is available in widths of 9, 12, 15 and 18 feet. The 9 and 12 foot widths are the most common in residential construction. Carpet is estimated in square yards. *Example,* a room 11'0" x 17'0" has 187 square feet. See Figure 15-1 for chart showing square feet in rooms. The carpet installer will probably plan to cut 17'0" from a 12'0" wide carpet roll to eliminate a seam. Therefore, he will estimate 204 square feet (12' x 17') which is 22.67 or 23 square yards. This is 2 square yards more than the room size. Check the room layout with the carpet installer to reduce waste to a minimum.

The owner will seldom have the carpet selected when the estimate is prepared. In your estimate allow so much per square yard installed or for all carpet in the house.

Resilient floor covering provides reasonable durability and economy of maintenance. It is installed over wood or concrete with adhesive. A floor underlayment over the subfloor is recommended for installation over wood.

Resilient tile (9" x 9" or 12" x 12") is available in asphalt, vinyl-asbestos, vinyl, rubber, cork and linoleum. Sheet goods (6' or 12' rolls) are available in vinyl and linoleum. Tile is sold by the square foot. Most suppliers do not break boxes, so you must make an allowance of 10% -20% for waste. Sheet goods are sold by the square yard. Discuss the room layout wtih the installer to determine the minimum number of seams and the least waste before preparing the estimate.

Other masonry floor coverings are used in better homes. Ceramic tile is used in bathrooms. Slate or other durable waterproof materials are used in foyers.

All floor covering is normally done on a contract basis by those who specialize in this work. As with carpet, the owner may or may not know the type and grade of floor covering he wants when the original estimate is made. In this case, make allowance for it.

Appliances

Built-in appliances such as the range, oven, range hood, dishwasher and garbage disposal are part of the building and should be included in the estimated cost of the house. Appliances such as the refrigerator, automatic washer and dryer may or may not be included in the estimate. Often the owner will agree to provide these appliances. If the make and model number of each appliance is not known, an allowance can be made for appliances that are listed on the specifications.

Labor

Some coordinating with the carpenters will be necessary. For example, after the carpet is installed, the interior doors will probably have to be cut off. Normally this is done by carpenters. The carpenters may be responsible for installing the built-in appliances, after which the electrician and plumber will make the necessary electrical and plumbing connections. Don't overlook this labor cost in your estimates.

The painting, floor covering and appliances for the house in Figures 7-1 and 7-2 are estimated as follows:

Painting

The interior and exterior painting, including all caulking, filling all nail holes and sanding was let on contract. The exterior was to have two finish coats of exterior latex paint over the prime coat. All interior gypsum wallboard was to be finished with two coats of latex flat paint except in the kitchen, bathrooms and utility room which were finished with a semigloss. All interior wood trim was to have three coats of latex paint with the final coat finished with semigloss. All woodwork was to have a smooth finish upon completion. The doors and trim in the family room were to have a natural finish. The contract price was $3200.00 for all labor and material.

There is no wall fabric to estimate.

Floor covering

Wall-to-wall carpet is to be installed throughout the house (including the main stairs) except the vestibule, kitchen, bathrooms and utility room. The carpet is to be installed over a rug pad with a weight not less than 80 ounces per square yard. $2500.00 was allowed for the carpet and installation.

The kitchen, bathrooms, vestibule and utility room are to have vinyl sheet goods. Make an allowance of $600.00 for this, including installation cost.

There are no other floor coverings in the living area of the house.

Appliances

The plans and specifications specify a range, oven, range hood, dishwasher and garbage disposal. The owner also wants to include a refrigerator in the allowance. Allow $1500.00 for these appliances.

Labor

The carpenters will have to cut off the interior doors after the carpet is installed. The electrical contract includes the installation and electrical connections for the built-in appliances. Estimate $100.00 for the labor for this coordinating work.

(Enter all of the above costs on the Cost Estimate Worksheet)

Cost Estimate Worksheet For Painting, Floor Covering And Appliances

Painting, labor and material	$________
Wallpaper, labor and material	________
Floor Covering:	
Carpet, installed	________
Resilient floor covering installed	________
Other, installed	________
Appliances	________
Labor for carpenters *	________
Total	$________

(Enter on Line 16, Form 100)

* Note: Coordinating work with the subcontractors.

Chapter 17

Gutters, Onsite Improvements and Miscellaneous

Gutters

Gutters provide controlled disposal of water from roofs and away from foundations to prevent damage to the property. They are sometimes omitted when roof overhangs exceed 12 inches for one story or 24 inches for two story houses.

Gutters may be galvanized steel, aluminum, copper, plastic, or wood. Seamless aluminum gutters are popular because they are prefinished and do not need painting. These gutters are formed by a machine on the job from roll aluminum and, as the name indicates, there are no seams except at corners and at the point of connection for the downspouts.

Downspouts, or leaders, should have one square inch of cross section for each 100 square feet of roof surface. The number and size of downspouts necessary for adequate disposal of the water can be calculated from this formula.

The installation of gutters is specialty work and should be done by specialists. The cost is normally based on linear feet for labor and material. To make an approximate estimate, the linear feet of the eaves plus the linear feet of the downspouts (each elbow and corner counts as one foot) for the total linear footage. This total multiplied by the cost per linear foot will give the estimator an approximate cost for the gutters. For a firm bid, consult the installer because there may be other factors that will increase the cost.

If drain pipe is not used to carry water away from the foundation to a point of discharge, splashblocks will be necessary. Splashblocks should be constructed of concrete or other durable materials with a minimum width of 12 inches and a minimum length of 30 inches. They should be firmly imbedded to prevent displacement.

If drain pipe is used from the downspouts at the foundation to the point of discharge, the labor and material cost per linear foot will be about the same as the cost of galvanized or aluminum gutter. But add the cost of the trench. For example, if the gutters cost $2.00 per linear foot and the trench costs $1.00 per linear foot to excavate, the drain pipe from the foundation to the point of discharge could be estimated at $3.00 per linear foot.

Onsite improvements

The specifications will indicate the dimensions of the driveway and the type of surface. If the driveway is over filled ground, asphalt or concrete should not be spread until the soil is compacted. The fill dirt may be compacted at the time it is placed, but the weight of vehicles, and the effects of rain, snow, freezing and thawing will cause additional settling for several months. Where the driveway enters the street a culvert pipe may be required. The building inspector will specify the size and type of culvert pipe that will be acceptable.

Other onsite improvements may be retaining walls, fences, barbecue pits, and accessory buildings. This additional work may be done on a cost plus basis or for a contract price. All required extra work should be detailed in writing, stating the method of payment.

Miscellaneous

Some of the many items that may not be listed on the specification sheet but must be included in the cost estimate are:

1. Additional lumber for scaffolding.
2. Polyethylene for ground covering in crawl spaces and to cover materials.
3. Equipment rental.
4. Additional nails, lag screws, bolts.
5. Weather stripping materials when not included elsewhere.
6. Access doors to crawl spaces, attics, etc.
7. Disappearing stairs.
8. Gravel for a temporary driveway during construction.
9. Miscellaneous hardware such as screen wire (where needed), corner braces for shelves, magnetic catches and hinges.
10. Miscellaneous molding such as screen mold.

11. Additional caulking.
12. Light bulbs during construction.
13. Railings not estimated elsewhere.
14. Drapery rods.
15. Undereave soffit vents, if not estimated elsewhere.
16. Special equipment such as attic fans not estimated elsewhere.
17. Cleaning materials and cleanup expense.

These are only some of the items that may be overlooked by the estimator. Some of them may be listed elsewhere in the bid, but they are worth noting. Their combined costs can be considerable.

The estimated cost of the gutters, onsite improvements and miscellaneous costs for the house in Figures 7-1 and 7-2 are:

Gutters

There are 301 linear feet of gutters and downspouts. The discharge is to be on splashblocks with an elbow at the bottom of each downspout. Prefinished white seamless aluminum gutters are to be used. The gutters were let on contract for $395.00 installed.

The owner wanted splashblocks in lieu of drainpipe. There are 8 - 12" x 30" precast concrete splashblocks to estimate.

Onsite improvements

The driveway from the double garage to the street is 16' x 77'. Because of the fill, no asphalt or concrete is to be estimated for the surface at this time. Crushed stone 4" thick is the only surface to estimate. The number of tons of crushed stone is estimated as follows:

$$\frac{16' \times 77' \times .33'\ (4")}{27} = 15.06 \text{ cubic yards}$$

Cubic yards x 1.35 factor = tons (see Figure 14-2)
15.06 cu. yds. x 1.35 factor = 20.33 or 21 tons

There will be a 12" x 20' culvert pipe under the driveway where it connects to the street. The materials for the driveway will be estimated at:

Order: *21 tons crushed stone*
1 - 12" x 20' corrugated culvert pipe

An allowance of $1000.00 is made for the landscaping. There are no other onsite improvements.

Miscellaneous

An allowance of $200.00 is made for gravel for the temporary driveway during the construction, additional nails and other hardware, cleaning materials and cleanup expenses.

Labor

The labor to spread the stone on the driveway, install the culvert pipe and other miscellaneous work is estimated at $200.00.

(Enter the above costs on the Cost Estimate Worksheet)

Cost Estimate Worksheet For Gutters, Onsite Improvements And Miscellaneous

Gutters $________

________ lin. ft. drain pipe @ ________ = $________

________ splash blocks @ ________ = ________

$________

Sales tax (____%) ________

________ ________

Onsite improvements:
Driveway:

________ tons stone @ ________ = ________

________ lin. ft. culvert pipe @ ________ = ________

________ sq. yds. asphalt @ ________ = ________

________ cu. yds. concrete @ ________ = ________

Other onsite improvements:
(List on separate sheet and enter total cost here) ________

$________

Sales tax (____%) ________

________ ________

Landscaping: ________

Miscellaneous: ________
(If necessary list on separate sheet and enter total cost here)

Labor: ________

Total $________

(Enter on Line 17, Form 100)

Chapter 18

Overhead, Contingency and Profit

Overhead

Many expenses on each job don't show up in plans and specifications. Some of these expenses are:

1. F.I.C.A. (Employer's share of Social Security taxes).
2. F.U.T.A. (Federal and state unemployment taxes paid by employer).
3. Workmen's Compensation insurance.
4. Liability insurance.
5. Administrative expenses to prepare weekly payrolls, and quarterly and end-of-year reports on employees.
6. Car and truck expenses.
7. Temporary telephone.
8. Interest on borrowed money.
9. Estimator's time.

If you are building on spec., you will have the overhead expenses listed above plus most of the following:

1. Fire and Liability insurance.
2. Taxes on the real estate.
3. Advertising.
4. Sales commission.
5. Legal fees.

Most of the overhead expenses will vary from job to job. Some builders add from 22% to 28% to the estimated payroll to cover these expenses. Others may prefer to list them separately under overhead. How they are charged to the job is not important. But any of these expenses overlooked reduce your profit.

Contingency
There are few houses constructed where the actual cost and the estimated cost are identical. Labor (unless it is contractor labor) is always an estimate. Inflation can increase material costs weekly. Costs that could not be foreseen when the estimate was prepared are another factor. However, the actual cost of the house should be within 5% of the estimated cost. To compensate for this difference, add an allowance for contingency. Experience from previous jobs is the best guide for the percentage to add. From 2% to 3% of the estimated cost of the house before profit is added should be an adequate allowance.

Profit
You are entitled to wages for your time on the job plus a profit on the money invested in your business. Your wages may be included in the overhead expense or in the profit. Profit often depends on the competition. Too much profit may result in losing the job. For new work, a profit of 8% to 15% is reasonable. The percent will vary with the amount of the bid, competition, and the estimated time you spend on the job.

Some states and cities have a business and occupation tax (B&O tax) that builders must pay on gross receipts. Where this tax exists, the builder is exempt from paying sales tax on the materials he uses. But he must pay the B&O tax on gross receipts. This tax, where applicable, must be added to the estimate after the profit is added.

The overhead, contingency and profit for the house in Figures 7-1 and 7-2 is estimated as follows:

Overhead
For all payroll taxes, insurance and overhead, 25% will be added to the estimated labor cost. The house is built on contract rather than on speculation.

Contingency
2% of the estimated cost of the house before the profit is added is allowed for the contingency.

Profit
10% of the total estimated cost of the house is added for the builder's wages and profit.

There are no state and city business and occupation taxes (B&O tax) where the house is built.

Form 100
Estimating Home Building Costs

Date____________________

1. The Building site $________
2. Preliminary costs ________
3. Site clearing, excavation and fill dirt ________
4. Footings ________
5. Foundation ________
6. Floor system ________
7. Superstructure ________
8. Roofing ________
9. Electrical, plumbing, heating and air conditioning ________
10. Brickwork ________
11. Energy saving materials ________
12. Interior wall and ceiling finish ________
13. Exterior trim ________
14. Concrete floors, walks and terrace ________
15. Interior trim ________
16. Painting, floor covering and appliances ________
17. Gutters, onsite improvements and miscellaneous ________
18. Overhead, contingency and profit:
    (a) Labor: $____ multiplied by ____ % = ________
    (b) Other: ________
    (c) Subtotal for lines 1 thru 18(b) $________
    (d) Contingency: (Line 18(c) multiplied by ____ %) = ________
    (e) Subtotal $________
    (f) Profit: (Line 18(e) multiplied by____ %) ________

Total $________

Glossary

A

Abrasive: Granular coating on sandpaper and grinding wheels used for smoothing.

Access door: A small door used for entry into attics and crawl spaces.

Accessory buildings: Small buildings used for storage and utility purposes.

Adhesive: A sticky or bonding substance used to hold materials together.

Aesthetics: Dealing with visual attractiveness.

Aggregate: Sand and stone added to cement to make concrete.

Air Entraining Agent: A chemical added to concrete which causes microscopic air bubbles for the purpose of resisting freezing.

Anchor bolts: Steel bolts embedded in concrete to tie the sill to the foundation.

Angle iron: A structural piece of steel shaped to form the letter "L" in a cross section.

Apron: Inside window trim placed against the wall under the stool.

Architect: One who is specially trained to design buildings.

Areaway: An enclosed space below grade adjacent to a basement window which allows ventilation and light to enter.

Arterial street: The principal street for through traffic.

Asphalt shingles: Shingles made from asphalt-impregnated felt and covered with mineral granules.

Astragal: A small molding shaped to form the letter "T" used between double doors.

Atrium: An open court within a building.

Attic: The space under a roof and above the ceiling of a house.

B

Backfill: Earth used to fill around foundation walls or as fill for excavation.

Baluster: One of a set of small vertical members that support the handrail of a balustrade for a stairway.

Balustrade: The complete set of balusters and handrail serving as an enclosure for stairways and balconies.

Baseboard: The finish board lining the plaster or gypsum wallboard where it meets the floor.

Basement: The lowest floor area in a building. Normally it is partially or entirely below ground level.

Baseshoe: A small molding between the floor and baseboard.

Batten: A narrow strip of wood nailed vertically over the joints of vertical siding.

Batter boards: Horizontal boards set to a predetermined elevation constructed at the corners of a proposed building from which lines are stretched to locate the outline of the foundation.

Beam: A horizontal structural member used to support loads.

Bearing: In architecture, the wall that supports a load such as a bearing wall. Also the portion of a piece of timber or steel which rests upon a wall.

Bench mark: A fixed point on some permanent object used as a reference in determining floor and grade elevations.

Bird's mouth: The cutout at the lower end of a rafter that rests on the wall plate.

Blueprints: An exact plan reproduced on sensitized paper in white lines on a blue background. Now more frequently done in blue lines on a white background.

Board foot (abbreviated bfm): A unit of measure for lumber. One bfm is 1 inch thick, 12 inches wide and 1 foot long.

Bottom chord: The lower horizontal member of a truss replacing the ceiling joist.

Brick: Masonry units made from baked or burned clay and molded into oblong blocks used in buildings and walls.

Bridging: Cross bracing between floor joists for reinforcement and distribution of floor loads.

B.T.U. (British Thermal Unit): One B.T.U. is the amount of heat required to raise the temperature of one pound of water 1 °F.

Builder: One who specializes in one type of construction such as residential building, as opposed to a general contractor who engages in many types of construction.

Building permit: An authorization from the local governing body to construct or remodel a building.

Building line: The line of the outside of a foundation.

Building site: The location of an actual or planned structure.

Built-up beam: Several framing members nailed together to act as one beam. *Example:* a 6" x 10" built-up beam is 3 - 2" x 10" nailed together.

Butts, door: Door hinges.

C

Calcium chloride: A white crystalline compound used in concrete to accelerate setting time and to retard freezing.

Carport: An open-sided shelter for an automobile.

Carriage (or stringer): The framing material that supports stair steps.

Casing: Finish trim around windows and doors.

Ceiling joists: Horizontal framing members resting on the wall plate to support the ceiling.

Cesspool: An underground catch basin to receive and retain sewage.

Chalking: Loose powder formed from a paint surface.

Chimney: Sometimes called a flue. A passage through which smoke or fumes escape from the furnace and/or fireplace.

Chimney saddle: A small sloping roof in back of a chimney, used to shed water.

Collar beam: A horizontal beam just below the ridge tying rafters together.

Collector street: A secondary street that serves as a feeder street.

Column post: A vertical shaft or pillar used to support loads.

Common rafter: A roof timber sloping up from the wall plate to the ridge.

Compaction: The process of applying pressure to loose material to remove air pockets and to make the mass denser.

Composition roofing: A term referring to roofing with asphalt materials.

Concrete: A hard compact substance made of sand, gravel, portland cement and water.

Condensation: The moisture produced when warm moist air comes in contact with a cold surface.

Conduit: A tube for electrical wires.

Contingency: Something likely to happen such as uncertain conditions and unforeseen costs.

Contingency fund: A sum of money set aside for unforeseen costs.

Contractor: An individual or company who contracts to do certain work for a stipulated sum.

Corner bead: A metal bead used on outside corners for plaster and gypsum boards.

Corner brace: Diagonal braces at the corners of stud walls.

Cornice: The part of the roof overhang that projects from the wall.

Cost-Plus Contract: Work done on actual cost plus a fixed fee or percent basis.

Counter flashing: A secondary and overlapping layer of flashing.

Crawl space: An unfinished area under a floor normally with little head room.

Cripple: A structural member that is cut less than full length.

Cripple rafter (or jack rafter): A rafter whose length is less than a common rafter. They run between the ridge and valley rafter, and between the wall plate and hip rafter.

Cripple stud: Studs used over door openings and over and under window openings.

Critical Path Method: A graphical method used to control the planning and scheduling of a construction job to minimize loss time.

Cubic foot: A unit of volume having three dimensions. One cubic foot equals one foot in length multiplied by one foot in width multiplied by one foot in depth.

Cubic yard: A unit of volume having three dimensions. One cubic yard equals three feet in length multiplied by three feet in width multiplied by three feet in depth, or 27 cubic feet.

Cul-De-Sac: A turnaround at the end of a dead end street.

Culvert: A pipe under a driveway or street used for drainage.

Cupola: A small dome or similar structure on a roof.

Custom-built: A house made to order and not factory-built.

D

Deck: Subfloors are often referred to as floor decks, and the roof sheathing is sometimes referred to as the roof deck.

Deed: A legal document showing transfer of property.

Detail: The separate items of a structure drawn for the purpose of clarity.

Dimension line: A line showing the distance between two points.

Disappearing stairs: Stairs that fold out of sight when not in use.

Door buck: The rough framing around the door opening in masonry or concrete walls to which the finish door frame is attached.

Door stop: A small molding around the door jambs and head against which the door closes. Also metal door stops attached to baseboards, door butts, doors and walls to protect walls and furniture from damage.

Dormer: A structure projecting from a sloping roof.

Downspout (or leader): The vertical pipe connected to the gutter for the purpose of discharging the roof water.

Drain tile: Pipe used to carry off water.

Drip edge (or drip cap): A metal strip placed over the roof sheathing along the cornice to allow the roof water to drip free into the gutters.

Drywall: Interior wall covering other than plaster.

E

Easement: The legal right a person has to use a designated portion of the land owned by another for the purpose of installing and maintaining underground utilities, or for use as a road.

Eaves: The lower portion of a roof that extends beyond the wall.

Efflorescence: Crystalline compounds appearing on masonry walls that change to a whitish powder when the water that carried them to the surface evaporates.

Elevation: Drawings of a section of a building made as though the observer was looking straight at it. Also used to denote grade and floor heights above or below a reference point such as a bench mark.

Estimator: One who prepares a bid or cost estimate on a construction project.

Excavation: Earth removal, normally for the purpose of constructing a dwelling or other structure.

Existing grade elevation: The grade elevation before any excavation or fill is done.

Expansion joint: A flexible joint used in concrete construction to allow thermal expansion and contraction, and thus prevent cracking.

F

Face brick: Brick of better appearance and quality used on the face of walls.

Factor: A number when multiplied by another number to form a product.

Factory-built house: A house built in component sections in a factory and transported to the job site.

Fascia: The vertical board on the end of rafters that are part of the cornice.

Felt: An asphalt impregnated rag fiber paper used under roofing.

Fenestration: The arrangement of windows in a building.

Fiberboard: Fibers of various substances pressed together to form sheets.

Fiberglass shingles: Shingles with an inorganic base composed of glass fiber mat and covered with mineral granules.

F.I.C.A. (Social Security): Federal Insurance Contribution Act. Known as FICA taxes paid by employer and employee.

Fill: Adding earth, etc. to an existing grade until a required elevation is reached.

Finish grade: The final grade elevation.

Fire brick: A brick used in fireplaces that is heat resistant.

Flashing: Sheet metal, or other material, over windows and doors and around chimneys to prevent water leakage into the building.

Flitch plate: A steel plate fastened with bolts joining wood timbers to form a beam.

Floor covering: Carpet, tile, sheet goods, etc. laid over a floor underlayment for the finish floor.

Flooring: A name normally applied to wood flooring laid over a subfloor.

Floor elevation: The height of a floor above or below a designated point such as a bench mark.

Floor joists: Framing timbers that support the floor.

Floor plan: A drawing showing room arrangements, door and window locations and all needed dimensions.

Floor system: That part of the floor that includes the sill plate, girder, floor joists, bridging and subfloor.

Floor underlayment: Materials, normally with a smooth surface such as particle board or plywood, that are laid over the subfloor to provide a smooth base for the carpet, tile, etc.

Flue: A passage in a chimney to allow the escape of smoke and fumes into the outer air.

Footing: A concrete base for a foundation wall, wider than the wall, designed to distribute the load to the soil.

Foundation: The supporting part of a structure, including the footing and the base course.

Framing: The wood skeleton of a building.

Frieze: A flat vertical board fastened to the wall.

Frost line: The depth to which frost will penetrate into the ground.

Furring strips: Wood strips used to level and receive finish surface material.

F.U.T.A. (FUTA tax): Federal Unemployment Tax Act. A tax paid by the employer on his employees.

G

Gable roof: A triangle roof sloping up from two walls.

Gambrel roof: A roof with two roof pitches from the eave to the ridge sloping up from two walls. The lower pitch is steeper than the upper pitch.

Girder: A large horizontal structural member used to support the ends of joists.

Glazing: Setting glass in windows, etc.

Grade: The surface of the ground, normally referred to around buildings.

Grade beam: A reinforced concrete beam poured over unstable soil to support loads.

Grantee: A person to whom a grant is made.

Grantor: A person who makes a grant.

Grounds: Wood strips used to control the thickness of plaster.

Gutters: A horizontal trough used to collect and carry off water

Gypsum board: A board with a gypsum core faced with paper used in drywall construction.

H

Handrail: A protective railing used on stairways.

Hasp: A locking device used on doors.

Head joint: The vertical joint between bricks or masonry blocks.

Header: A structural member which supports the ends of the joists. It is also the name used for beams over window and door openings.

Hearth: The masonry floor in front of a fireplace.

Heat pump: A unit used for both heating and cooling.

Hip rafter: A diagonal rafter at the junction of two sloping roofs running from the plate to the ridge to form a hip.

Hip roof: A roof sloping up from all walls of a building.

Hose bibb: A water faucet used for the attachment of a hose.

House package: A term used to indicate all items purchased in a factory-built house.

I

I-Beam: A steel beam shaped in the form of the letter "I".

Infiltration: Air seepage around windows and doors.

Insulation: Special materials used over ceilings, under floors and in walls to retard transfer of heat.

Invert: The lowest point in a pipe or ditch.

J

Jack rafter: A rafter whose length is less than a common rafter. It runs between the ridge and valley rafter, and between the wall plate and hip rafter.

Jamb: The vertical member of a door or window frame.

Joist: One of the framing members that support the ceiling or floor.

Joist hanger: A metal strap or hanger to support one end of a joist.

K

Kick plate: A metal plate attached to the bottom of a door used to protect it.

Kiln-Dried lumber: Lumber dried in artificial heat such as in a kiln.

Knee wall: A low wall normally required in 1½ story houses.

L

Lally column: A steel column pipe filled with concrete for added strength.

Laminated beam: A beam glued together in layers under pressure.

Leader (or downspout): See Downspout

Ledger strip: A wood strip nailed to the bottom of a girder or beam on which notched floor joists are attached.

Lightweight concrete: Concrete made with an aggregate lighter in weight than sand. Vermiculite, Perlite and Pumice are lightweight aggregates.

Linear foot: One foot in length.

Lintel: A horizontal support of wood, steel or concrete across the head of a door or window opening.

Lookout: A short horizontal framing member that extends from the wall to the rafter that supports the finish soffit material.

Louver: An opening that provides ventilation while providing protection from rain and insects.

Lumber: Timber sawed in a sawmill such as boards and framing members.

M

Manhole: An opening to allow access to a sewer line.

Mansard roof: A roof with two roof pitches sloping up from the eave to the ridge on all four sides. The lower pitch is steeper than the upper pitch.

Mantel: The facing and/or shelf about a fireplace.

Masonry: A term used for building material such as brick, concreteblock, stone, etc. bonded with mortar.

Masonry cement: A mixture of cement and hydrated lime.

MBM: Thousand (feet) board measure.

Meeting rail: The horizontal center rails of a window.

Merchant builder: One who builds houses to sell on the open market.

Metes and Bounds: A system used in land surveying that describes the direction and distances of property lines until the perimeter has been traced around to the starting point.

Mil: Used to measure the diameter of wire and the thickness of paint. 1 mil = .001 inch.

Miter: The beveled surface cut on the end of casing and molding.

Modular house: A house built and finished in a factory and transported to the job site in two or more sections to be erected on a foundation.

Molding: A strip of material used for decoration.

Mortar: Masonry cement, sand and water used to bond brick, concrete blocks, stone, etc. together.

Mullion: A vertical bar between two windows.

Muntin: Small bars separating glass panes in a window.

Mural: A large picture on wall fabric or painted on a wall.

Muriatic acid: Hydrochloric acid used to clean bricks.

N

Newel post: The post at the head or foot of stairs supporting the handrail.

Nominal size: The size of lumber before it is dressed.

Norman brick: Brick with dimensions of 2¼" x 3¾" x 11½", or thereabouts.

Nosing: The rounded edges of a stair tread that extend beyond the riser.

O

On center (o.c.): Measurement from center-to-center of framing members.

Outlet: The point where a lighting fixture, heater, appliance or other current-consuming device is attached to a wiring system.

Overhang: The projection of one part of a structure over another, such as a roof overhang.

Overhead: Expenses charged to a construction project that are not directly a part of the construction costs, such as administrative expenses.

P

Paneling: A term used for finishing walls with wood panels.

Parapet: A low wall around the edge of a roof.

Parging: A cement coating applied to masonry walls.

Parquet floor: Flooring laid in geometric patterns.

Particle board: Wood chips bonded together with heat and pressure to form sheets.

Penny (d): A term indicating the length of nails such as 8d (2½''), and 10d (3'').

Percolation test: A test to determine the ability of soil to absorb water, normally used before a permit is issued to construct a septic tank.

Perpendicular: A line at right angles to another line.

Pier: A masonry pillar to support the floor framing.

Pilaster: A pier attached to a wall used for the purpose of strengthening the wall.

Piling: Wood or concrete posts driven down to a solid base in the earth to provide safe footing.

Plank: A long, broad, thick board at least 1½'' thick.

Planning Commission: A commission appointed by the local governing body to prepare a comprehensive plan as a general guide for the development of the area, and as a basis for the preparation of zoning and other regulations. The commission has the power of enforcement as prescribed by law.

Plat: A drawing, map or plan of a piece of land.

Plate: A horizontal framing member such as sill plate, sole plate and top plate.

Plot plan: A drawing showing the description of the lot, location and dimensions of the house, garage, walks and driveway, easements, utility lines, and finish floor and grade elevations.

Plumb: A wall that is in true vertical alignment.

Plywood: Wood built up from veneer sheets glued together under pressure with their grains at right angles to one another.

Polyethylene film: A lightweight thermoplastic film that is resistant to chemicals and moisture. It is primarily used as a vapor barrier in construction work.

Porch: A covered entrance to a building that projects from the wall and has a separate roof.

Portland cement: A cement that derives its name from a patent that was taken out in 1824 in England for the manufacture of an improved cement, which, when hardened produced a yellowish-gray mass resembling the stone found in various quarries on the Isle of Portland, England. This cement is the most widely used in concrete construction.

Precast concrete: Reinforced concrete structural units manufactured in a plant and transported to the job site.

Prefab fireplace: A fireplace built in a factory and transported to the job site.

Prefab house: See Factory-built house.

Prime coat: The first coat of paint applied to wood or metal.

Proportions: The relation between things with respect to size, amount, quantity, etc.

Purlin: A structural member laid over trusses to support rafters.

PVC: Polyvinyl Chloride Pipe used for water lines, waste lines, etc.

Q

Quarry: A place where stone is excavated.

Quoins: Exterior corners of a building distinguished from the adjoining surfaces by color, size or projection.

R

Radii: The plural of radius. A straight line from the center to the periphery of a circle.

Rafter: A diagonal framing member supporting a roof.

Rafter supports: Framing members inserted (usually diagonally) under rafters to strengthen the roof.

Railing: Horizontal top member of a balustrade.

Rake: The slanting edge of a gable roof.

Rake board: The vertical board along a rake.

Ratio: A proportion of fixed relation between two similar things.

Ready-mix concrete: Concrete mixed in a concrete plant and transported to the job site in specially designed trucks.

Realty: Real Estate.

Reinforced concrete: Concrete with steel rods or wire mesh inserted to increase its tensile strength.

Reinforcing rods: Steel rods used to reinforce concrete.

Renovation: A term used to describe remodeling a building.

Retaining wall: A wall that retains earth, used to eliminate steep banks.

Ridge: The highest point on a roof where two slopes meet.

Rise: The vertical height of a roof or stairs.

Riser: The vertical member of stairs between treads.

Roman brick: Bricks with dimensions of 1½" x 3¾" x 11½", or thereabouts.

Roof: The top covering of a building made of joists, trusses, or rafters, roof sheathing and roofing.

Roofing: Waterproof roof covering.

Roof pitch (or roof slope): The steepness of a roof measured in the number of inches of rise per 12 inches of run. A roof with a pitch of 6 in 12 will rise 6 inches for every 12 inches of run.

Roof underlayment: See Felt.

Roof ventilating louvers: Louvers placed on top of a roof for ventilation.

Rough grade: Shaping of the ground to an approximate elevation and contour.

Rough opening: An unfinished opening in a building.

Rowlock: Brick laid on edge.

Run: One half the span of a roof. The horizontal distance of a flight of stairs.

R-Value: The resistance value of insulating material measured by its ability to resist heat transfer. R-Values are numbered: The higher the number, the more effective the insulation.

S

Sash: A frame for holding the glass of a window or door.

Scaffold: A temporary wooden or metal frame for supporting workmen and materials.

SCR brick: Brick with dimensions of 2¼" x 5½" x 11½", or thereabouts.

Screeds: A wooden or metal form used as a thickness and leveling guide for concrete.

Scribing: Marking and fitting a piece of lumber to an irregular surface.

Septic tank: A private sewerage disposal system where waste matter is collected in a tank, putrefied and decomposed through bacterial action, and the liquid discharged into a disposal bed where it is absorbed into the soil.

Sheathing: The rough covering over the frame of a house.

Shed roof: A roof with only one slope. Normally built against a higher wall.

Shim: A thin wedge-shaped piece of wood or metal used for filling and leveling.

Shingles: A term used for pieces of material used for covering roofs and the sides of houses, such as asphalt shingles.

Shoring: Timbers used for temporary support.

Shutters: A movable louvered cover for a window.

Siding: The boarding on the outside of a house.

Sill: A horizontal wood or masonry unit under a window or door. The horizontal member bolted to the foundation wall called the mudsill or sill plate.

Sill plate (or Mudsill): See Sill.

Slab: A term used for a concrete floor.

Soffit: The underside of an overhang:

Solar: Something coming from the sun such as heat and energy.

Soldier course: Bricks set in a vertical position.

Sole plate: The horizontal framing member under a stud wall.

Span: Distance between two supports.

Specifications: Written instructions that describe the type and quality of material that is not shown or specified on the blueprints. The specifications become part of the blueprints and are referred to as the "plans and specifications" in a contract.

Speculative builder: See Merchant builder.

Splashblock: A masonry unit placed under a downspout to disperse the water away from the foundation.

Square: The material needed to cover 100 square feet.

Standard brick: Brick with dimensions of 2¼" x 3¾" x 8", or thereabouts.

Starter course: The roof shingles or felt laid around the eave and under the first course of roofing.

Stile: The vertical piece of a window or door frame.

Stool: A wood shelf across the bottom and inside of a window.

Stops: See Door stops.

Storm window: A window used with a regular window for added protection against the cold weather.

Stretcher course: Brick laid lengthwise so the side only appears on the face of a wall.

Stringer: See Carriage.

Struck joint: A finished mortar joint.

Struts: The vertical or inclined members between the top chord and bottom chord of a truss.

Stucco: A cement coating for walls.

Stud: The vertical framing member of a wall.

Sub-contractor: One who contracts part of a job from the principal contractor, such as an electrical contractor or a masonry contractor.

Sub-floor: Sheathing nailed to the floor joists to receive the finish floor.

Sump pump: A pump used to move water from a lower to a higher elevation.

Superstructure: That part of a building above the foundation.

Surfaced lumber: Lumber dressed in a planing mill designated as S4S, etc.

Symbol: A mark, sign or object used to represent another thing.

T

Terrace: An unroofed, paved area adjacent to a house.

Terrain: The natural contour of land.

Thermal glass: Two or more glass panes sealed together in a factory and separated by hermetically sealed air spaces.

Thimble: A horizontal pipe running through the chimney wall into the flue.

Tie wire: Wire used to tie reinforcing rods together.

Toe nail: Nailing at an angle.

Ton of Refrigeration: 12,000 BTU's

Top chord: The upper horizontal framing member of a truss which is the nailing base for the roof sheathing.

Top plate: The upper horizontal framing member of a stud wall.

Trapezoid: A quadrilateral of which only two sides are parallel.

Tread: The horizontal surface of a step.

Trim: The finish carpentry work in a house.

Trimmer: Joist or rafter around an opening in the floor or roof into which a header is framed.

Truss: A single framework consisting of top chord, bottom chord and struts joined together for supporting a roof.

Turnout: The lower end of a handrail on a balustrade that makes a spiral or twisting turn.

U

Underpinning: A support under a wall already built.

Undressed lumber: Rough lumber without a smooth finish.

Utilities: Water, sewer, electric, gas, telephone, TV cable, etc.

V

Valley: The bottom intersection of two roof slopes.

Valley jacks: Rafters between the ridge and valley rafter.

Valley rafter: The diagonal rafter under a valley.

Vapor barrier: Waterproof membrane such as Polyethylene film used to prevent passage of moisture.

Variance: A deviation from a zoning ordinance such as noncompliance with the required setback regulations. A waiver may be granted by the local Planning Commission for some variances.

Vent: A small opening to permit the escape of gas and fumes.

Vertex: The highest point of two intersecting lines furthest from the base.

Vestibule: A small entrance hall or room.

Volume: The cubic contents of the space occupied by three dimensions.

Volute: See Turnout.

W

Walers: Supports for concrete forms.

Wallpaper (or Wallfabric): Decorative paper or fabric for covering walls and ceilings.

Wall plate: See Top plate.

Wall ties: Corrugated metal strips used to tie brick veneer walls to the adjacent wall.

Wainscot: The lower part of a room with a different finish from the upper, such as paneling on the lower wall.

Waterproofing: Preventing the entrance of water.

Weather stripping: A strip of fabric or metal around doors and windows to prevent air leakage.

Weeping tile: See Drain tile.

Winders: A stair tread wider on one end than the other.

Window stool: See Stool.

Wire mesh: Wires welded together in rectangular or square grids used to reinforce concrete.

Workmanship: The skill of a worker.

Z

Zoning ordinance: An ordinance adopted by the local governing body to protect the health, safety, property and welfare of the public. All building construction is regulated by the zoning ordinances.

Index